T0210896

Communication and Control for Networked Complex Systems

Chen Peng · Dong Yue
Qing-Long Han

Communication and Control for Networked Complex Systems

 Springer

Chen Peng
School of Mechatronic Engineering
 and Automation
Shanghai University
Shanghai
China

Qing-Long Han
Griffith School of Engineering
Griffith University
Gold Coast, QLD
Australia

Dong Yue
Research Institute of Advanced Technology
Nanjing University of Posts and
 Telecommunications
Nanjing
China

ISBN 978-3-662-52632-3   ISBN 978-3-662-46813-5   (eBook)
DOI 10.1007/978-3-662-46813-5

Springer Heidelberg New York Dordrecht London
© Springer-Verlag Berlin Heidelberg 2015
Softcover reprint of the hardcover 1st edition 2015

Printed on acid-free paper

Springer-Verlag GmbH Berlin Heidelberg is part of Springer Science+Business Media
(www.springer.com)

Preface

Networked control systems (NCSs) are control systems where the control loops are closed via communication channels. With the rapid development of networking communication technologies, NCSs have received increasing attention in the past decade. Compared with traditional point-to-point control systems, NCSs have a number of advantages, such as low cost, easy maintenance, and increased system flexibility. More recently, much attention has also been paid to design the suitable communication scheme to save the limited communication resource for NCSs.

The study of NCSs can be twofold: designing controller over a preselected network and designing appropriate communication scheme for satisfying the requirement of control. In the first designing controller over a preselected network field, Internet Protocol (IP) networks are commonly preselected in NCSs due to their well-developed infrastructure, wide acceptance, simplicity, and affordability. The IP-based network delays display a nonuniform distribution character with small delays being dominant while large delays being exiguous, which implies that the probability of small delays is bigger than the one of large delays. However, this feature has not been well used in the analysis and synthesis of NCSs. Therefore, how to effectively use nonuniform delay distribution character in NCSs has stimulated the first research line of this monograph.

In the second designing appropriate communication scheme field, communication bandwidth is a scarce resource in NCSs. If signal transmission only occurs when the relevant information from the sensor-to-controller needs to be transmitted, then not only all of the transmitted signals from the controller-to-actuator are helpful to ensure the desired performance, but also the redundant information is possibly avoided in the transmission. Therefore, more limited network resources can be released to other communication tasks in need. Notice that time-triggered communication schemes are a common assumption in some existing results, which leads to inefficient utilization of the limited network resources. Therefore, how to design efficient communication schemes to save the limited communication resource while ensuring the desired performance of NCSs has stimulated the second research line of this monograph.

Structure and Readership

This book is structured into three parts; Part I is devoted to introduce an overview of recent developments of NCSs (Chap. 1) and provide a summary of the modeling, communication schemes, and mathematical lemmas used in the derivation of the main results of this book (Chap. 2); they are the premises of the following two parts. Part II is devoted to consider the nonuniform distribution communication character of IP-based communication networks in the analysis and synthesis of linear, nonlinear and large-scale systems under network environments; Part III is devoted to design communication schemes to save the limited network resources while ensuring the desired performance, that is, discrete event-triggered communication schemes, self-triggered communication schemes, co-design of communication and control considering the data loss and communication delay in the communication, and mixed self and event-triggered communication scheme for improving the energy efficiency.

Part II: Internet Protocol (IP) networks are generally used in NCS. Network delays display irregular behavior in Internet Protocol (IP) networks. However, this feature has not been well explored in some existing results in NCSs. Therefore, in Chap. 3, a networked delay distribution-dependent H_∞ control for networked linear control systems is proposed. In Chap. 4, delay nonuniform distribution character of network delay is considered in the PDC fuzzy controller design. Moreover, a premise reconstruct method is also proposed to deal with the asynchronous premise problem of T-S fuzzy systems in network environments. A decentralized control method is proposed for networked large-scale systems considering the above-mentioned delay distribution dependence in Chap. 5.

Part III: Most works on NCSs so far assume that the sampled data by the sensors are periodic transmitted over the communication networks, i.e., time-triggered communications. In general, a time-triggered communication scheme leads to inefficient utilization of limited network resources. Therefore, In Chap. 6, an adaptive event-triggered communication scheme is provided, which can dynamically adjust the event-triggered communication threshold to reduce the conservativeness induced by time-invariant communication threshold. A co-design method to consider the event-triggered communication and robust H_∞ control in a unified framework is presented in Chap. 7, allowing part of event-triggered packets which can be lost in communication. For saving the limited energy in wireless NCSs, in Chap. 8, a novel self-triggered sampling scheme is proposed for the execution of sampling in NCSs by taking into consideration network-induced delays and data packet dropouts. In Chap. 9, a mixed sampling scheme for the execution of sampling in wireless NCSs is proposed by striking a balance between self-triggered sampling and event-triggered sampling to achieve high energy efficiency. In Chap. 10, a discrete event-triggered communication scheme is proposed for a networked T-S fuzzy system, which can reduce unnecessary communication while ensuring the desired control performance.

Acknowledgments

We would like to acknowledge the collaborations with Prof. Yuchu Tian and Dr. Engang Tian on the work of communication delay distribution and Prof. Minrui Fei and Dr. Taicheng Yang on the work of event-triggered communication reported in the monograph and Ph.D. candidates: Jin Zhang, Qiang Zheng, and Heng Zhang for their great help in this monograph. The supports from the National Natural Science Foundation of China under Grant (61273114), the Innovation Program of Shanghai Municipal Education Commission under Grant (14ZZ087), the Pujiang Talent Plan of Shanghai City, China under Grant (14PJ1403800), the International Corporation Project of Shanghai Science and Technology Commission under Grant (14510722500), and the Natural Science Foundation of Jiangsu Province of China under Grant (BK20131403) are gratefully acknowledged. Finally, the close cooperation with Springer as publisher and particularly with Dr. Lu Yang as responsible editor is gratefully acknowledged.

Shanghai, China, January 2015   Chen Peng
Nanjing, China   Dong Yue
Gold Coast, Australia   Qing-Long Han

Contents

Acronyms

A	System matrix
A^{-1}	Inverse of matrix A
A^T	Transpose of matrix A
$A \geq 0$	Symmetric positive semi-definite
$A > 0$	Symmetric positive definite
$A \leq 0$	Symmetric negative semidefinite
$A < 0$	Symmetric negative definite
$det(A)$	Determinant of matrix A
$diag(X_1, X_2, \ldots, X_m)$	Diagonal matrix with X_i as its ith diagonal element
I	Identity matrix of appropriate dimensions
lim	Limit
LMI	Linear matrix inequality
$\mathbb{N}$	Positive integers
$Ones(m, n)$	A $m \times n$ matrix of ones
$Prob(\cdot)$	Probability
$\mathbb{R}$	Field of real numbers
$\mathbb{R}^n$	n-dimensional real Euclidean space
$\mathbb{R}^{n \times m}$	Space of $n \times m$ real matrices
$sgn(x)$	The sign of x
$tr(A)$	Trace of matrix A
h	The sampling interval of sensor
$0_{n \times m}$	Zero matrix of dimension $n \times m$
$\lambda(A)$	Eigenvalue of matrix A
$\lambda_{\min}(A)$	Minimum eigenvalue of matrix A
$\lambda_{\max}(A)$	Maximum eigenvalue of matrix A
$x(k)$	The state variable vector at time kT
$|x|$	Absolute value (or modulus) of x
$\|x\|$	Euclidean norm
$\|P\|$	Induced norm $sup_{\|x\|=1} \|Px\|$
$\forall$	For all

$\in$	Belong to
$\rightarrow$	Tend to, or mapping to (case sensitive)
$\otimes$	Matrix Kronecker product
$\sum$	Sum
$\mathbb{E}\{\cdot\}$	Mathematical expectation operator
sup	Supremum
inf	Infimum
$*$	Entries implied by symmetry

Part I
Introduction and Preliminaries for NCSs

Chapter 1
Introduction

1.1 What Are Networked Control Systems?

With the rapid development of computer and network technologies, conventional control systems have been evolving to modern networked control systems (NCSs), which implement control functionality over communication networks. They are becoming increasingly important in industrial process control due to their cost-effectiveness, reduced weight and power requirements, simple installation and maintenance, and high reliability. Introductions to NCSs can be found in survey papers [1–4] and references therein.

An NCS integrates information, communications, and control into a single system in which control loops are closed over communication networks. A frame diagram of an NCS is depicted in Fig. 1.1, where the communication network may be a wired network, a wireless network, an IP-based network, and so on. Compared with conventional point-to-point control systems, the introduction of communication network in the closed-loop system results in some major difficulties in NCS design and implementation. A typical packet transmission processing in an NCS is shown in Fig. 1.2, in which three typical cases are described as follows:

- **Network-induced delays** affect the accuracy of timing-dependent computing and can degrade the control performance significantly. Generally, the network-induced delays have time-varying characteristics. They may degrade the control performance and introduce distortion of controller signals.
- **Data dropouts** result from network traffic congestions and limited network reliability. When a data packet is dropped, complete information of the NCS becomes unavailable. In this case, the controller or actuator has to work with incomplete information.
- **Disorder packets** mean that earlier transmitted packet arrives at the destination later, which implies that the latest available sampling or control packets at the controller and the actuator may be not the latest transmitted packets. In this case, the controller or actuator has to decide or control with uncertainty information.

© Springer-Verlag Berlin Heidelberg 2015

C. Peng et al., *Communication and Control for Networked Complex Systems,*
DOI 10.1007/978-3-662-46813-5_1

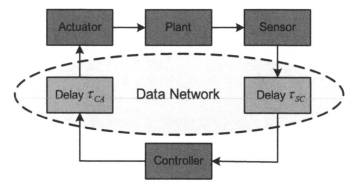

Fig. 1.1 A frame diagram of NCSs

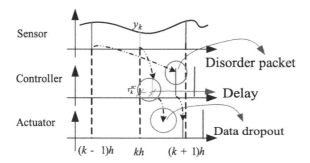

Fig. 1.2 A typical packet transmission processing in NCSs

Besides the points mentioned above, the configuration or deployment of NCSs, such as, the multi-packet transmission [5, 6], the signal quantization [7–9], distributed sensors and controllers [10, 11], may also affect the performance of NCSs. Moreover, the above-mentioned problems in the deployment of NCSs become more evident and severer when wireless networks are employed and/or when the limited computing and network resources are considered. Therefore, a general assumption in traditional control theory, i.e., all communication links that are necessary for solving a control problem are instantaneous, data lossless and reliability are not satisfied in NCSs. This means that many existing traditional control technologies may become infeasible for specific networked control applications. Stability analysis and control design of NCSs has to take into account communication constraints with respect to network-induced delays, data dropouts, variable communication topology, and so on.

1.2 Control of NCSs

During the last decade, different control methods have been developed for the control
of NCSs with consideration of network-induced delays, data dropouts, communication bandwidth constraints, and so on. Some of the typical control methods in NCSs
are described as follows:

1.2.1 Delay-Dependent Control in NCSs

For NCSs, time-delay modeling-based control methods are typically introduced in
[12, 13]. In [12], for a system controlled over a communication network, under the
assumption that the state of the system is available for measurement and there is a
zero-order hold (ZOH) between the controller and the actuator, a linear state feedback
controller is adopted and the closed-loop system is described by

$$\dot{x}(t) = Ax(t) + BKx(t_kh), \quad t \in [t_kh + \tau_{t_k}, t_{k+1}h + \tau_{t_{k+1}}) \tag{1.1}$$

where h is the sampling period, t_k $(k = 1, 2, 3, \ldots)$ are some integers and $\{i_1, i_2,$
$i_3, \ldots\} \subset \{0, 1, 2, \ldots\}$, τ_{t_k} is the time delay from the sensor to the actuator. It is not
required that $t_{k+1} > t_k$. If $\{i_1, i_2, i_3, \ldots\} = \{0, 1, 2, \ldots\}$, it implies that no packet
dropout occurs in the transmission. When $t_{k+1} = t_k + 1$, it means that $h + \tau_{t_{k+1}} > \tau_{t_k}$,
which includes $\tau_{t_k} = \tau_0$ and $\tau_{t_k} < h$ as special cases.

Define $\tau(t) =: t - t_kh$ for $t \in [t_kh + \tau_{t_k}, t_{k+1}h + \tau_{t_{k+1}})$. Then the system (1.1) is
evolved as

$$\dot{x}(t) = Ax(t) + BKx(t - \tau(t)), \quad t \in [t_kh + \tau_{t_k}, t_{k+1}h + \tau_{t_{k+1}}) \tag{1.2}$$

Based on the above-mentioned modeling idea, further works can be seen in [14] to
deal with output feedback case for an NCS with the input/output signal quantization,
and in [15] to consider the nonlinear uncertainties for an NCS.

From the definition of $\tau(t)$, it is clear for $t \in [t_kh + \tau_{t_k}, t_{k+1}h + \tau_{t_{k+1}})$

$$\dot{\tau}(t) = 1, \quad \tau_{t_k} \leq \tau(t) \leq (t_{k+1} - t_k)h + \tau_{t_{k+1}} \tag{1.3}$$

which is shown in Fig. 1.3, where the artificial delay $\tau(t)$ has a sawtooth structure with
lower and upper bounds, this property is used in [16] to derive the less conservative
results and schedule the allowable number $|t_{k+1} - t_k|$ of data dropouts, sampling
period h, network-induced delay $\tau_{t_{k+1}}$.

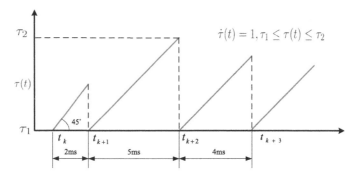

Fig. 1.3 Piecewise time-varying delay $\tau(t)$ in an NCS

1.2.2 Stochastic Control in NCSs

A Markov chain is a stochastic process, which refers to the sequence of random variables such a process moves through, with the Markov property defining serial dependence only between adjacent periods. Therefore, it can be used for describing the actions of an NCS that follow a chain of linked events, such as, network-induced delays or data dropouts, where what happens next depends only on the current state of the studied system [17, 18]. For example, a Markov packet dropout model is depicted in Fig. 1.4, where the packet transition probabilities are given by

$$\begin{cases} Pr\{d(k+1) = 0|d(k) = 0\} = q \\ Pr\{d(k+1) = 1|d(k) = 0\} = 1 - q \\ Pr\{d(k+1) = 1|d(k) = 1\} = p \\ Pr\{d(k+1) = 0|d(k) = 1\} = 1 - p \end{cases} \tag{1.4}$$

The main advantages of the Markov model are: (i) the dependencies between delays are taken into account since in real networks the current time delays are usually related with the previous delays, and (ii) the packet dropout could be included naturally [19]. For example, in [20], an optimal stochastic control approach is presented for the control of a system over a random delay network. The effects of

Fig. 1.4 Markov packet
dropout model

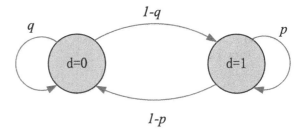

network delays are treated as a linear quadratic gaussian problem. It requires that the sum of the sensor-to-controller and the controller-to-actuator time delay is less than the sampling period, and information of all the past delays are available. In [21], the stabilization problem is addressed for an NCS with Markovian characterization, the closed-loop system is modeled as a Markovian jump linear system with two jumping parameters. More recently, by taking the full advantage of the packet-based transmission in NCSs, a delay compensation control approach is proposed in [22] to actively compensate the network-induced delay in a Markovian jump linear system framework.

This stochastic control approach has shown better application in the analysis and synthesis of NCSs. However, if the network-induced delays or the data dropouts in the communications do not satisfy the Markov chain assumption, this model and control method will be infeasible in NCSs.

1.2.3 Switched Control in NCSs

A switched control system is a hybrid dynamical system composing of a finite number of subsystems, where the subsystems are described by differential or difference equations, and a logical rule orchestrates switching between these subsystems [23]. A survey of basic problems in stability analysis and stabilization of switched systems can be seen in [24]. Applying the switched control method in NCSs is a topic of significant interest [25, 26], since there are solid theoretical results existing in the literature for stability analysis and control design for switched control systems [27, 28].

The basic idea of switched control in NCSs is to formulate an NCS as a discrete-time switched system. A conceptual framework of switching control in an NCS is depicted in Fig. 1.5, where an NCS is modeled as a discrete-time switched control

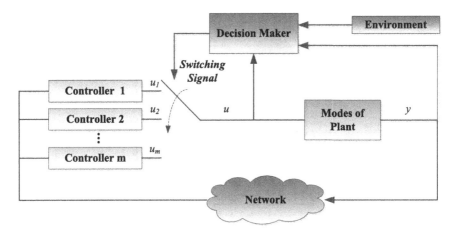

Fig. 1.5 A conceptual framework of switching control in an NCS

system, decision-maker decides which controller to be used based on the current modes of the system and exoteric environments.

Then stability and synthesis problems of NCSs can be reduced to corresponding problems for the discrete-time switched systems. There are some results available in the literature [25, 29–31]. For example, in [29], an NCS is modeled as a discrete-time switched linear uncertain system, which allows the controller to be given in discrete time as well as in continuous time. In [30], under uncertain access delay and packet dropout effects, stability and disturbance attenuation issues for a class of NCSs are considered in the framework of switched systems. In [25], under consideration of the data packet dropouts in both the forward and the backward communication channels, an NCS is modeled as a discrete-time switched time system with a time-driven observer-based output feedback controller.

Notice that the switch control generally requires that the controller works at a higher frequency than the sampling frequency. Furthermore, how to consider both the communication channels from the sensor to the controller and the controller to the actuator, and how to unify model delay and data dropout together can be further studied.

1.2.4 Predictive Control in NCSs

The predictive control is used in NCSs to overcome the negative effects of network-induced delay and data loss in the communications [32]. A framework of predictive control of NCSs is shown in Fig. 1.6, where the network delay compensator is used to compensate the network-induced delays and data losses in the forward (from controller-to-actuator) and feedback (from sensor-to-controller) channels and achieves the desired control performance.

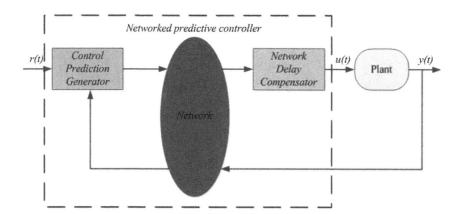

Fig. 1.6 Predictive control of NCSs

Assume that a set of data can be transmitted at the same time through a communication network, the predicted sequences are therefore sent to the actuator in a package. In [33], the network delay compensator is used to choose the control value from the latest prediction sequence as the control input value. Similar control method is also adopted in [34] under consideration of network-induced delay with the Markov chain characteristic. Moreover, for wireless network control, a method is proposed in [35] to codesign of predictive controllers; a predictive compensation method is proposed in [36] under consideration of the network delays and packet losses in NCSs. Considering the packet loss in NCSs, a model predictive control method is also used in [37] to study the stabilization of linear systems over networks with bounded packet loss, an input-to-state stability of packetized predictive control of NCSs is addressed in [38].

Predictive control has the novel advantage in dealing with network-induced delay and data loss in NCSs. However, predictive control depends on the system model and the current and the past states to predict the future state, therefore, it is a model-based method in nature. If the model is not known a priori and there are larger disturbances and uncertainties in the studied system, how to deal with these cases is still challenging.

1.2.5 Model-Based Control in NCSs

To see how infrequent feedback information is needed to guarantee that the system remains stable, a model-based control method for NCSs is proposed in [39, 40]. The conceptual framework of model-based control is depicted in Fig. 1.7, where the plant model is used at the controller side to approximate the plant behavior during the interval of transmitted data arriving at the controller.

The main idea of model-based control for NCSs is to perform the feedback by updating the model's state using the actual state of the plant provided by the sensor.

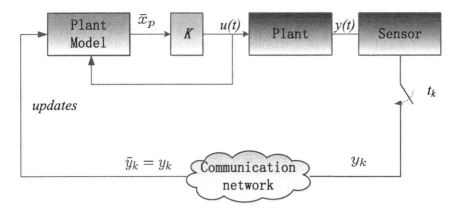

Fig. 1.7 A conceptual framework of model-based control in an NCS

Different from the data coming from the network are held constant between sampling times in [12, 13], the transmitted data is used to instantaneously update the state of the controller in [39, 40].

Notice that the model-based control idea is to switch between the open-loop control and closed-loop control, which has the advantage of significantly increasing the periods where the system is operating in an open-loop fashion. Moreover, since only a communication channel between the sensor and controller is considered in [39, 40], as an extentation, both the communication channels from the sensor-to-controller and the controller-to-actuator can be further studied for model-based NCSs.

1.2.6 Fuzzy Logic Control in NCSs

Fuzzy control method has proven to be a successful control approach to many complex nonlinear systems or even nonanalytic systems, which has also been suggested as an alternative approach to conventional control techniques in many cases [41]. Moreover, as a further and significant step in the system synthesis of T-S fuzzy systems, the parallel distribution compensation (PDC) [42] control structure utilizes a nonlinear state feedback controller, which mirrors or shares the structure of the associated T-S fuzzy model.

Under the PDC framework, there are some results in the literature on the T-S fuzzy systems controlled though a communication network in the last few years. For example, identical timescale premises and asynchronous lagged premises were adopted in [43–45] and [46–48], respectively. However, since the introduction of network environments, the fuzzy control rules cannot fully share in the T-S fuzzy system by making use of the traditional PDC strategy [49, 50]. For example, in [48], the inferred fuzzy controller is given by $u(t) = \sum_{j=1}^{r} h_j(\theta(t_k))K_j x(t_k)$, and the resulting closed-loop system is expressed as

$$\dot{x}(t) = \sum_{i=1}^{r}\sum_{j=1}^{r} h_i(\theta(t))h_j(\theta(t_k))[A_i x(k) + B_i K_j x(t_k) + \cdots] \qquad (1.5)$$

where $t \in [t_k h + \tau_k, t_{k+1}h + \tau_{k+1})$, τ_k is the communication delay, and "$+ \cdots$" denotes the other terms. Hence, it is known that $h_i(\theta(t)) \neq h_i(\theta(t_k))$ in (1.5) due to the fact that there exist communication delays between the plant and the controller. As a result, due to the different $h_i(\theta(t))$, $h_i(\theta(t_k))$ in (1.5), the method in traditional point-to-point T-S system to decompose the accumulative items into more separate items to reduce the conservativeness is infeasible in networked T-S fuzzy systems. In this case, the obtained controller gains are identical, which shows that a PDC controller does not offer any advantages over a linear controller for the studied system controlled over a communication network [50, 51].

To widen the applicability of the fuzzy control method under network environments and reduce the conservativeness, a method to reconstruct the grades of membership at the controller is proposed in [51], which guarantees that grades of membership at the fuzzy plant and at the fuzzy controller have the same timescale. As a result, not only the problem of leading/lagging utilization of the latest information in the premises of PDC fuzzy rules in some existing ones is avoided, but also the conservativeness induced by the asynchronous premise is reduced. In [52], the deviations between $h_i(\theta(t))$, $t \in [t_k h + \tau_k, t_{k+1} h + \tau_{k+1})$ and $h_j(\theta(t_k))$ is addressed for networked T-S fuzzy systems with asynchronous normalized membership functions, and a method to determine the deviation bounds is also established.

1.3 Communication of NCSs

In the development of NCSs, time-triggered and event-triggered communication (ETC) schemes are commonly adopted sampling schemes. For time-triggered communications, the system's states are sampled periodically and all sampled signals are transmitted with the same sampling frequency as the sensor. The advantages of periodic sampling procedures are that the well-developed theory on sampled-data control systems may be used in the analysis and synthesis of NCSs. However, from a resource allocation point of view, if all sampled-data are transmitted over the communication networks, then unnecessary utilization of the limited network resources occurs. A general diagram of a time-triggered communication scheme is depicted in Fig. 1.8. One can see from Fig. 1.8 that all sampled signals are transmitted over

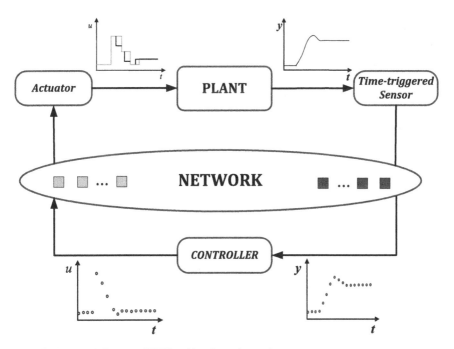

Fig. 1.8 A general diagram of NCSs with a time-triggered sensor

Fig. 1.9 An illustration of
event-triggered sample

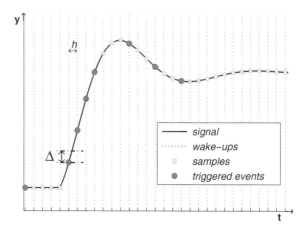

the communication networks, therefore, this periodic time-triggered communication
scheme can lead to significant overprovisioning of NCSs [53].

For ETC, the signal is transmitted after the occurrence of an external event, generated by a predesigned event-triggering condition, rather than the elapse of times as
in conventional periodic time-triggered communication. A general diagram of ETC
scheme is depicted in Fig. 1.9. One can see from Fig. 1.9 that not all sampled signals
are transmitted over the communication networks, only those satisfying a preselected
threshold condition have the right to be transmitted over a communication network
[53]. Since ETC has the advantage in maintaining the control performance while
saving limited communication resources, therefore, there has been growing interest
in ETC [10, 54, 55]. Moreover, there are some aperiodic communication methods
in the literature. For example, it is assumed in [56] that a node does not broadcast
the true value of the state of its local process except that it differs from the estimate known to the remaining nodes by more than a given threshold. In [57, 58],
the trade-off between the amount of information exchanged and the performance is
investigated, and a stochastic communication logic for an estimator-based NCS is
proposed to reduce the communication load. Under consideration of the stabilization of a group of linear systems whose feedback loops are closed through a shared
network with limited bandwidth, a feedback communication policy is proposed in
[55] for deciding which system should be admitted into the networks. These works
can be seen as ETC scheme since the transmission events are not triggered except
that they are truly needed in NCSs.

1.3.1 Continuous Event-Triggered Communication Scheme

Based on the available information for measurement, there are two kinds of ETC
schemes, that is, state-dependent communication scheme [59, 60] and output-
dependent communication schemes [61–63]. The communication tasks are executed
only if the predesigned state/output-dependent triggering conditions are violated.

Define the error $e(t)$ as

$$e(t) \triangleq x(t) - x(t_k), \quad t \in [t_k, t_{k+1}) \qquad (1.6)$$

where t_k is the latest instant when the sensor transmits the sampled data to the controller, $x(t_k)$ and $x(t)$ are two system states at the latest transmitted sampling instant and the current instant, respectively. Assume that $e(t)$ in (1.6) can be online calculated based on real-time measured state $x(t)$, the following state-dependent ETC scheme is used to determine whether the current sampled data should be transmitted or not in [10, 59],

$$t_{k+1} = t_k + \min\{t | g(e(t), x(t_k)) \geq \delta\} \qquad (1.7)$$

where δ is a preselected constant threshold, $g(\cdot)$ is a predefined function related with $e(t)$ and $x(t_k)$. Compared with time-triggered communication scheme, the communication frequency can be effectively reduced based on (1.7). Notice that the scheme (1.7) is dependent on continuous state $x(t)$, where the special hardware is necessary for successive measurement and computation [64]. As an extendibility, if the state is not available for measurement in (1.6), output-based ETC schemes are provided in [61–63] to save the limited communication resources while ensuring the desired performance. Moreover, to avoid Zeno behavior [63], it is required that the nonzero lower bound of sampling and communication intervals in (1.7).

1.3.2 Discrete Event-Triggered Communication Scheme

Consider the discrete sampling character of NCSs, earlier discrete ETC schemes schemes are proposed in [65, 66], and further works can be found in [63, 67–69]. Discrete ETC schemes mean that only discrete measured state is used to determine whether or not the measured state should be transmitted, which has the advantage in saving the limited network resources, while preserving the desired performance and without resorting to extra hardware.

Although the ETC schemes in [65, 66] have the different threshold conditions, the key idea of schemes mentioned above can be described as

$$t_{k+1} = t_k + \min_{l \in \mathbb{N}}\{lh | f(e(t_k + lh), x(t_k)) \geq \delta\} \qquad (1.8)$$

where $e(t_k + lh) := x(t_k + lh) - x(t_k)$, h is a constant sampling period, $l = 1, 2, \ldots$, $f(\cdot)$ is a predefined function related with $e(t_k + lh)$ and $x(t_k)$.

From (1.8), one can see that not all of the measured states are transmitted through the communication network, that is, only the value of $f(e(t_k + lh), x(t_k))$ at the discrete sampling instants $t_k + lh$ violates the prescribed threshold δ, then the measured state is transmitted over the communication networks. Compared with those continuous event-triggered schemes in [10, 59], where the extra hardware is necessary to fulfill the continuous measurements and computation [10, 70], the extra hardware for continuous measurement is not necessary. Moreover, different from those in [10] through simulations to show that the transmission periods are greater than a positive constant as the state goes to the equilibrium, the discrete communication scheme (1.8) ensures that the lower bound of the transmission period is a constant sampling period h.

1.3.3 Asynchronous Event-Triggered Communication Scheme

For a distributed system with the system's components located in different places with large scales and high dimension, such as a distributed NCS as depicted in Fig. 1.10, where ETM is the abbreviation for "event-triggering mechanism." It is difficult to use the traditional centralized ETC scheme mentioned in above Sects. 1.3.1 and 1.3.2 to collect all the sampled data together to decide whether to transmit them or not.

The general expression of distributed ETC schemes can be written as

$$t_{k+1}^i = t_k^i + \min_t \{t \,|\, f_i(e_i(t), x_i(t_k)) \geq \delta_i\} \tag{1.9}$$

where i means the number of subsystems, $f_i(\cdot)$ and δ_i are the triggering functions and triggering thresholds of the subsystems i.

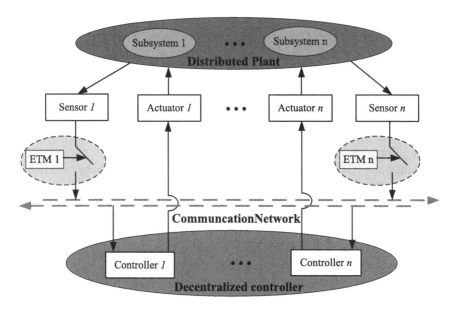

Fig. 1.10 A distributed NCS with distributed ETC scheme

There are some results on distributed ETC schemes in the literature [10, 71–73]. For example, for a large class of nonlinear related systems, a distributed synchronous ETC scheme is proposed in [71], while one subsystem meets distributed transfer conditions, all event generators send the current sampling data. For a distributed nonlinear NCS, an ETC scheme is proposed in [10], where a subsystem broadcasts its state information to its neighbors only when the subsystem's local state error exceeds a specified threshold. If the controller is not known a priori, the aforementioned methods are no longer valid again. Moreover, in the analysis and synthesis of distributed NCSs, how to deal with the asynchronous triggering instants t_{k+1}^i and how to design the triggered functions $f_i(\cdot)$ and thresholds δ_i are still challenging.

1.3.4 Self-triggered Communication Scheme

The self-triggered communication (STC) scheme is a software implementation of an ETC scheme. Compared with a passive ETC scheme, which determines whether an transmitted event is triggered or not based on a online calculated or estimated triggering condition, a STC scheme can predict the next trigger time by the current available information to guarantee the desired control performance, while using less communication resources and energy [64].

The idea of STC scheme is to predict the time it takes for $\|e(t)\|$ to go from $\|e(t_k)\|$ to δ, i.e.,

$$t_{k+1} = \max\{t \mid \|e(t)\| = p(t, x(t), x(t_k)) \le \delta, t \in [t_k, t_{k+1})\} \qquad (1.10)$$

where δ is a pre-given scalar, $p(\cdot)$ is a forecasting function related with $e(t)$ and $x(t_k)$. In this way, the next maximum allowable transmitted instant t_{k+1} can be estimated by bounding the sampling error $\|e(t)\|$ through δ.

Following the idea of self-triggering control scheme presented in [74], an energy-efficient self-triggered sampling scheme for an NCSs controlled over IEEE 802.15.4 networks is provided in [75]. The ETC and STC schemes over sensor/actuator networks are analyzed in [76], a self-triggered coordination of robotic networks for optimal deployment is proposed in [77]. Inspired by a "self-triggered" task model in a real-time control system [78], a STC scheme is designed in [79] for an NCSs under consideration of communication delays and data dropouts simultaneously.

Compared with ETC scheme, STC scheme has the advantage in saving the energy, especially in wireless sensor networks, the demand for continuous supervision and measurement in the sensor leads to additional power consumption [80, 81]. However, if there are state-independent disturbance in the studied system, communication delays and data dropouts in the communications, one must strive to design more appropriate $p(\cdot)$ in (1.10).

1.4 Contribution of the Book

In this book, our focus is how to use the IP-based communication delay character in the analysis and synthesis of NCSs, and how to save the communication resources while ensuring the desired performance. Therefore, communication delay distribution-dependent analysis of synthesis methods for networked linear, nonlinear, and large-scale systems are the first contribution of the book; a series of communication schemes considering the communication resource constraints and system character are the second contribution of the book. More precisely, the main contributions of this book are generalized as follows:

- Typical statistical characteristics of communication delays are analyzed for a class of IP-based communication networks. IP-based communication delay is shown to be nonuniformly distributed with short delays occurring in a high probability and long delays in a low probability. To fully use above delay distribution character, a new stochastic NCS model is proposed to capture the nonuniform distribution of the IP-based communication delays.
- For NCSs, a stochastic delay distribution-based stability analysis and controller design approach is proposed, it builds a bridge to connect Quality of Control (QoC) with network Quality of Service (QoS). For stability of NCSs, statistical delay distribution-dependent stability criteria are derived in the form of LMIs, in which a new tighter bounding technique for cross terms is involved and improved results are obtained. Because of the use of the delay distribution information, much less conservative results are obtained than those from existing results. Since the obtained conditions for the existence of admissible controllers are not expressed as strict LMIs, the cone complementary linearisation procedure is employed to solve the non-convex feasibility problem.
- To reduce the conservativeness induced by the asynchronous premises in T-S fuzzy systems under network environments, a method to reconstruct the grades of membership at the controller is proposed, which guarantees that grades of membership at the T-S fuzzy plant and at the fuzzy control rules have the same timescale. As a result, both the problem of leading/lagging utilization of the latest information in the premises of parallel distributed compensation (PDC) fuzzy rules in some existing ones and the conservativeness induced by the asynchronous premise are tackled. Combining the above asynchronous premise reconstruct method, a networked T-S fuzzy modeling approach is also developed to capture the nonuniform distribution and multifractal nature of the IP-based communication delay, which ensures that the network-induced delays and data losses can be modeled in a unified framework.
- A decentralized control framework for a large-scale system controlled over an IP-based communication network is proposed. Compared with the centralized control scheme adopted in the literature, the requirement of signal packet transmission among the distributed subsystems is no longer required; and the analysis and synthesis criteria for the system under consideration capture the nonuniform distribution characteristic of an IP communication network and depend on the

topology of the communication networks. The decentralized control scheme with the identical or different controller gains is decided by whether or not there is a communication between the controllers and actuators.

- A discrete event-triggered transmission scheme to determine whether or not the sampled-data should be transmitted is proposed. The idea of this scheme is that: (i) the sampled-data error between the sampled-data at the current sampling instant and the sampled-data at the last transmission instant is first calculated, and (ii) when a specified threshold is violated, the sampled-data is transmitted. Not like some existing event-triggering conditions need to be monitored continuously, here the condition is only checked at each sampling instance, the transmission period in the proposed scheme is time varying with multiple sampling periods. Therefore, the special hardware for real-time measurement and calculation is no longer needed. Moreover, a codesign algorithm is provided to obtain the parameters of the event-triggered transmission scheme and the controller gain simultaneously.

- To relax the assumption in some existing ones that ETC threshold is a preselected constant, a novel adaptive ETC scheme is proposed. The release times of the sampled data are triggered by an adjustable threshold, which can be adaptively adjusted based on the latest available information. As a result, the number of transmitted packets through the communication networks is reduced, and the desired control performance is achieved while consuming the fewer network resources. Moreover, combining the proposed discrete ETC scheme together, a PDC fuzzy controller with asynchronous premise constraints is deliberately designed. Hence, the controller must be prior known in some existing works is no longer required in this book.

- A novel self-triggered sampling scheme for an NCS is developed by taking into consideration network-induced delays and data dropouts. The next sampling instant is predicted as a function of previously sampled-data, where the next sampling instant of the sensor is predicted based on the desired performance and the latest accepted time-stamped control packets; and network-induced delays and data dropouts are considered simultaneously in the design of self-triggered sampling scheme. Compared with some existing ones, the proposed self-triggered sampling scheme has the advantage in reducing the communication load, while preserving the desired control performance.

- A mixed self- and event-triggered (MSE) sampling scheme for a wireless NCS is proposed to reduce the energy expenditure in terms of the number of transmitted packets and idle listening period, while ensuring the desired control performance. The main advantage of the proposed MSE with respect to some existing ETC/STC schemes is that: the next sampling instant is predicted based on STC scheme to extend the idle listening period of WSNs, while avoiding the Zeno behavior and infinity sampling; and following the above-determined communication instant, a backstepping error dependent ETC scheme is used to determine whether the current sampled data should to be transmitted or not to reduce the conservativeness induced by SET, while avoiding the time mismatch problem. As a result, the conservativeness induced by the STC scheme is also reduced due to the proposed ETC component in MSE. Moreover, it is no longer needed to keep the radio on for all the time to periodically check the event-triggered condition due to STC scheme component in MSE. Therefore, the proposed MSE captures both the advantages of STC scheme and ETC to reduce the communication load and improve the energy efficiency, while preserving the desired performance.

1.5 Book Outline

The book is organized as follows: Part I consists of Chaps. 1 and 2, which is devoted to introduce an overview of recent developments of NCSs and provide a summary of the modeling, communication schemes, and mathematical lemmas used in the derivation of the main results of this book. Part II consists of Chaps. 3–5, addressing the nonuniform distribution communication character of IP-based communication networks in the analysis and synthesis of linear, nonlinear, and large-scale systems under network environments. Part III includes Chaps. 6–10, devoting to design series of communication schemes to save the limited network resources, while ensuring the desired performance.

In Chap. 2, the communication delay distributed-dependent model of NCSs and series of ETC schemes to be used in Parts II and III are introduced.

In Chap. 3, networked delay distribution-dependent state feedback control and H_∞ control for networked linear control systems are proposed.

In Chap. 4, delay nonuniform distribution character of network delay is considered in the fuzzy controller design. Moreover, a premise reconstruct method is also proposed to deal with the asynchronous premise problem of T-S fuzzy systems in network environments.

In Chap. 5, a decentralized control method is proposed for networked large-scale systems considering above-mentioned delay distribution-dependent character.

In Chap. 6, an adaptive ETC scheme is provided for networked H_∞ filtering design, which can dynamically adjust the ETC threshold to overcome the conservativeness induced by time-invariant communication threshold.

In Chap. 7, a codesign method to consider the ETC and robust H_∞ control in a unified framework is presented, which allows part of event-triggered packets can be lost in the communication.

In Chap. 8, for saving the limited energy in wireless NCSs, a self-triggered sampling scheme is proposed for the execution of sampling by taking into consideration network-induced delays and data dropouts simultaneously.

In Chap. 9, a mixed sampling scheme for the execution of sampling in wireless NCSs is proposed by striking a balance between self-triggered sampling and event-triggered sampling to achieve high-energy efficiency.

In Chap. 10, a discrete ETC scheme is proposed for a networked T-S fuzzy system, which can overcome the unnecessary communication, while ensuring the desired control performance.

Chapter 2
Preliminaries: Modeling, Communication Scheme, and Lemmas for NCSs

In this chapter, delay distribution-dependent modeling and discrete event-triggered communication (ETC) scheme are introduced for NCSs studied in this book. The main idea of delay distribution-dependent modeling is to use the nonuniform distribution character of IP-based networks in the modeling of NCSs, therefore, it is helpful to reduce the conservativeness in the analysis and synthesis of NCSs. Compared with the time-triggered communication scheme, the proposed ETC scheme has the advantage in preserving the desired control performance, while saving the limited network bandwidth.

2.1 Delay Distribution-Dependent Modeling

2.1.1 Nonuniform Distribution of IP-based Network Delays

IP-based networks, such as, Internet networks, Ethernet networks, and IP-based wireless sensor networks are generally used in NCSs. Actual IP-based network delays are time-varying and display irregular behavior. To show the characteristics of actual communication delays in IP-based networks, the Round-Trip Time (RTT) delays from different Ethernet network nodes are measured in [82]. Their results have showed nonuniform distribution of the RTT delays in the IP-based networks. For an NCS over 10 Mbps IP networks, NCSs over IP networks are simulated in [83] by making use of the open source package NS2. Part of the delays from the sensor-to-controller over 240 s are shown in Fig. 2.1. The corresponding histograms of the delays are shown in Fig. 2.2. From Figs. 2.1 and 2.2, one can see that the IP-based communication delay is an approximate of the probability density.

Moreover, Fig. 2.2 shows that the histograms of the delays skew to the left, which indicates that the probability of the large delays are shorter than the probability of the median and mean delays. Actual delays can be much longer than the median and mean, but with much lower probability. Therefore, communication delay is

© Springer-Verlag Berlin Heidelberg 2015
C. Peng et al., *Communication and Control for Networked Complex Systems*,
DOI 10.1007/978-3-662-46813-5_2

Fig. 2.1 IP-based network delays from the sensor-to-controller

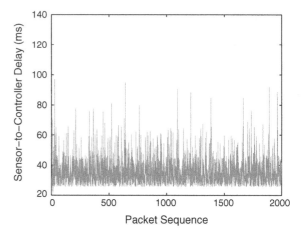

Fig. 2.2 Percentage of the IP-based network delays shown in Fig. 2.1

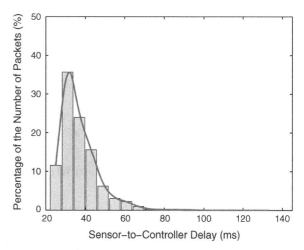

time-varying, nonuniform, and has multifractal nature [82, 84, 85]. Furthermore, accumulative percentage of IP-based network delays is shown in Fig. 2.3.

Compared with the upper bound of delay τ_3, the middle delay τ_2 reflects more information on the communication network, and $\tau_3 - \tau_2$ is far larger than $\tau_2 - \tau_1$, where τ_1 is lower bound of delay in Fig. 2.3, which also show the nonuniform distribution of the network-induced delay [85]. In the analysis and synthesis of NCSs, if the information of statistical distribution of IP-based network delays is fully utilized, less conservative results are expected to be achieved [86].

For well use of aforementioned nonuniform time-varying communication delays in the analysis and synthesis in NCSs, in the following, a delay distribution-dependent modeling approach is proposed to handle this scenario.

Assume that the network-induced delay $\tau(t)$ is approximately continuous with lower and upper bounds and can be formulated as

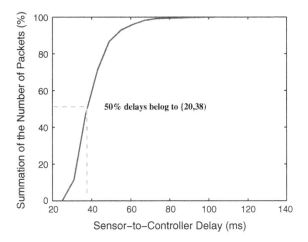

Fig. 2.3 Accumulative percentage of the network-induced delays shown in Fig. 2.2

$$\tau_1 \leq \tau(t) \leq \tau_3 < \infty, \quad \forall t \geq 0 \tag{2.1}$$

From the above discussion, it is known that small delays are dominant while large delays are exiguous. Therefore, the probability of small communication delay $\tau(t)$ is bigger than that of large delays. From the control application point of view, this phenomenon can be described using the following conditions:

$$\begin{cases} (1)\, \tau_1 \leq \tau(t) \leq \tau_3 < \infty, \quad \forall t \geq 0 \\ (2)\, Prob[\tau(t) \in [\tau_1, \tau_2)] = \bar{\delta}, Prob[\tau(t) \in [\tau_2, \tau_3]] = 1 - \bar{\delta} \end{cases} \tag{2.2}$$

where $\bar{\delta} \in [0, 1]$, τ_2 is the bound which means the probability of $\tau(t) \in [\tau_1, \tau_2)$ equals to $\bar{\delta}$.

Define a stochastic variable $\delta(t)$ as

$$\delta(t) = \begin{cases} 1, & t \in \Omega_1 \\ 0, & t \in \Omega_2 \end{cases} \tag{2.3}$$

where $\Omega_1 = \{t : \tau(t) \in [\tau_1, \tau_2)\}$, $\Omega_2 = \{t : \tau(t) \in [\tau_2, \tau_3]\}$. It is clear that $\Omega_1 \cup \Omega_2 = \mathbb{R}^+$ and $\Omega_1 \cap \Omega_2 = 0$. Based on the definitions of Ω_1 and Ω_2, $t \in \Omega_1$ means the event $\tau(t) \in [\tau_1, \tau_2)$ occurs and $t \in \Omega_2$ means the event $\tau(t) \in [\tau_2, \tau_3]$ occurs.

In this book, the Bernoulli distributed white sequence is used to describe the network quality of service (QoS) related stochastic variable $\delta(t)$.

$$\begin{cases} Prob\{\delta(t) = 1\} = \mathbb{E}\{\delta(t)\} := \bar{\delta} \\ Prob\{\delta(t) = 0\} = 1 - \mathbb{E}\{\delta(t)\} := 1 - \bar{\delta} \end{cases} \tag{2.4}$$

The above-described statistical distribution of IP-based network delay is one of the key characteristics of NCS network QoS. The fundamental difficulty from this new description of network QoS is how to improve the QoC under this new framework. In the following section, a class of system controlled over an IP-based network is used to show how to use this nonuniform delay distribution character in NCSs.

2.1.2 A Delay Distribution-Dependent Control Model for NCSs

Consider a class of control systems governed by

$$\dot{x}(t) = g(x(t), u(t), \omega(t)) \tag{2.5}$$

$$x(t) = \rho(t), \quad t \in [-\tau_3, 0] \tag{2.6}$$

where $g(\cdot)$ is a function related with $x(t)$, $u(t)$ and $\omega(t)$; $x(t) \in \mathbb{R}^n$ and $u(t) \in \mathbb{R}^m$ are state vector and control input vector, respectively; $\omega(t) \in \mathcal{L}_2[0, \infty)$ denotes the exogenous disturbance signal; $\rho(t)$ is the given function which means the initial condition of the systems on the segment of $t \in [-\tau_3, 0]$.

To make use of the delay distribution information, a delay-dependent feedback control law is designed as:

$$u(t) = \delta(t)Kx(t - \tau_1(t)) + (1 - \delta(t))Kx(t - \tau_2(t)) \tag{2.7}$$

where K is the feedback gain of the network controller and

$$\tau_1(t) = \delta(t)\tau(t), \tau_2(t) = (1 - \delta(t))\tau(t) \tag{2.8}$$

and $\delta(t)$ as described in (2.3).

Substituting (2.7) into (2.5), the following model for the closed-loop control systems can be obtained:

$$\begin{cases} \dot{x}(t) = g(x(t), \delta(t)Kx(t - \tau_1(t)) + (1 - \delta(t))Kx(t - \tau_2(t)), \omega(t)) \\ x(t) = \rho(t), \quad t \in [-\tau_3, 0] \end{cases} \tag{2.9}$$

Notice that two delay regions and white sequence $\delta(t)$ with Bernoulli distribution are considered in (2.9). If more information is available on the network QoS, the results developed in this section can be easily extended to a more general form

$$\dot{x}(t) = g(x(t), \sum_{i=1}^{N} \delta_i(t)Kx(t - \tau_i(t)), \omega(t)) \tag{2.10}$$

where N is the number of delay classification, $\delta_i(t)$ are stochastic variables which satisfying $Prob\{\delta_i(t) = 1\} = \mathbb{E}\{\delta_i(t)\} := \bar{\delta}_i$, $\sum_{i=1}^{N} \bar{\delta}_i = 1$. $\delta_i(t)$ and $\tau_i(t)$ can be obtained through a formula similar to (2.8).

Moreover, when $\bar{\delta} \triangleq 0$ or 1 and lower delay bound τ_1 is not considered, model (2.9) is reduced to the following special case:

$$\dot{x}(t) = g(x(t), Kx(t - \tau(t)), \omega(t)) \tag{2.11}$$

$$x(t) = \rho(t), \quad t \in [-\tau_3, 0] \tag{2.12}$$

This implies that the statistical distribution characteristics of network communication delay $\tau(t)$ is not considered in system analysis and synthesis. This special case has been investigated in [12, 15, 26].

The delay-dependent modeling method mentioned above will be used in Chaps. 3–5 to deal with linear systems, nonlinear systems and large-scale systems, respectively.

2.2 Event-Triggered Transmission Scheme

This section introduces series of ETC schemes applied in this book to reduce the number of transmitted data. Firstly, two discrete ETC schemes are presented with/without considering the data dropouts, respectively. Secondly, an adaptive event-triggered communication scheme is provided to dynamically adjust the event-triggered threshold parameter to reduce more numbers of transmitted packets, while ensuring the desired control performance. Finally, a completed NCS model is provided to link the ETC scheme with the other part of the system to be controlled.

To further development, the following assumptions are needed:

Assumption 1 The sensor is time-triggered with a constant sampling period h. The sampling sequence is described by the set $\mathbb{S}_1 = \{0, h, 2h, \ldots, kh\}$, where $k \in \mathbb{N}$.

Assumption 2 Whether the sampled data should be transmitted or not over a communication network is determined by a predetermined communication scheme. The transmission sequence at the sensors is described by the set $\mathbb{S}_2 = \{0, b_1h, b_2h, \ldots, b_kh\} \subseteq \mathbb{S}_1$, where $b_k \in \mathbb{N}$. Moreover, part of the data in $\mathbb{S}_2$ may be lost in the communication.

Assumption 3 The controllers and the actuators are event-triggered. The successfully transmitted sampled sequence at the sensors is described by the set $\mathbb{S}_3 = \{0, t_1h, t_2h, \ldots, t_kh\} \subseteq \mathbb{S}_2$, where $t_k \in \mathbb{N}$.

Assumption 4 The control input at the actuator is generated by a zero-order holder (ZOH) with the holding time $t \in \Omega \triangleq [t_kh + \tau_{t_k}, t_{k+1}h + \tau_{t_{k+1}})$, where τ_{t_k} is the communication delay, h is the sampling period, and $t_kh + \tau_{t_k}$ are the instants when the control signal reaches the ZOH.

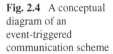

Fig. 2.4 A conceptual diagram of an event-triggered communication scheme

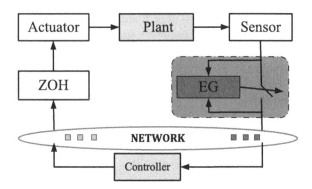

2.2.1 A Discrete Event-Triggered Communication Scheme

The conceptual diagram of an ETC scheme adopted in this book is shown in Fig. 2.4, where an event generator (EG) is located between the sensor and controller. The EG detects the sampled-data error between the sampled data at the current sampling instant and the sampled data at the latest transmission instant. Whether the sampled data should be transmitted or not is decided by a preselected threshold. The sampled-data transmission should be executed only when the specified threshold is violated [87]. Assume that there is communication delay in the communications but no packet loss is considered, the next transmission instant in the event-triggered transmission scheme is designed as

$$t_{k+1}h = t_kh + \min_{\ell}\{\ell h \ \big| \ e^T(i_kh)\Phi e(i_kh) \geq \delta_1 x^T(t_kh)\Phi x(t_kh)\} \qquad (2.13)$$

where $\delta_1 \geq 0$ is a given scalar parameter, Φ is a positive definite weighting matrix to be designed, and $e(i_kh)$ is the error between the two states at the current sampling instant and the latest transmitted sampling instant, i.e.,

$$e(i_kh) = x(i_kh) - x(t_kh) \qquad (2.14)$$

where $i_kh = t_kh + \ell h$, $\ell \in \mathbb{N}$; h is the sampling period of the sensor; t_k ($k=1, 2, 3, \ldots$) are some integers such that $\{t_1, t_2, t_3, \ldots\} \subset \{0, 1, 2, 3, \ldots\}$; t_kh is the time at that instant a packet is successfully transmitted from the sensor. The weighting matrix $\Phi > 0$ in (2.13) can also be adjusted to meet the requirement of the event-triggered transmission scheme.

Notice that $e^T(i_kh)\Phi e(i_kh) < \delta_1 x^T(t_kh)\Phi x(t_kh)$ implies that no sampled data is transmitted; $e^T(i_kh)\Phi e(i_kh) \geq \delta_1 x^T(t_kh)\Phi x(t_kh)$ means that a transmission event is generated, and the ZOH uses the latest sampled data as the input of the actuator and holds it until the next transmitted sampled data.

From (2.13), for $t \in [t_kh + \tau_k, t_{k+1}h + \tau_{k+1})$, the following inequality is satisfied. This will be used in the derivation of the main results in Chaps. 6–8.

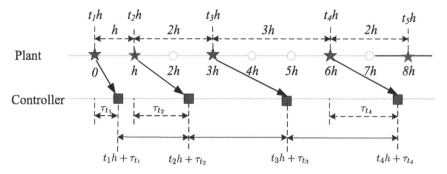

Fig. 2.5 An example of time evolution of the sampling and transmission series

$$e^T(i_k h)\Phi e(i_k h) < \delta_1 x^T(t_k h)\Phi x(t_k h) \tag{2.15}$$

Notice that the communication scheme (2.13) is characterized by the parameters δ_1, Φ and h, which determine the communication load. How to design these parameters to reduce the amount of data transfer and at the same time to meet the performance requirement is studied in [69]. As a special case, if $\delta_1 \to 0+$ in (2.13), this leads to $t_{k+1}h = t_k h + h$, which means that all the sampled data should be transmitted, the event-triggered transmission scheme (2.13) is simplified as a time-triggered transmission scheme in [88, 89].

In Fig. 2.5, an example is used to show a typical time evolution with a periodic sampling and an event-triggered transmission scheme. One can see that not all the sampled data is transmitted, only the sampled data satisfying the condition in (2.13) is transmitted. For example, the sampled data at the instants 0, h, $3h$ and $6h$ is transmitted, and the sampled data at other instants is not transmitted.

With the communication scheme (2.13), if there is no packet loss, the transmitter can directly utilize the violation of (2.13) to determine to send packets, while preserving the desired performance of the system under consideration. With packet loss, (2.13) cannot be directly employed to determine if the sampled data should be transmitted. In [69], (2.13) is changed to:

$$b_{k+1}h = b_k h + \min_{v}\{vh \,\big|\, \tilde{e}^T(l_k h)\Phi\tilde{e}(l_k h) > \delta_2 x^T(b_k h)\Phi x(b_k h)\} \tag{2.16}$$

where $\tilde{e}(l_k h) = x(l_k h) - x(b_k h)$, $l_k h = b_k h + vh$, $v \in \mathbb{N}$, $\delta_2 > 0$ is a given scalar parameter, $\Phi > 0$ as defined in (2.13), $b_k h$ is the transmitted sampling instant as defined in Assumption 2.

Notice that the state error and $l_k h$ in (2.16) are different from those in (2.13), and δ_2 in (2.16) should be less than or equal to δ_1 in (2.13) to take into account the extra communication delay caused by the data dropout.

The communication scheme (2.16) is characterized by the parameters δ_2, Φ and h, which determine the communication load. Also, if $\delta_2 \to 0+$ in (2.16), this leads to $t_{k+1}h = t_kh + h$, and this becomes a time-triggered sampling scheme in [88, 89]. How to ensure the allowable number of data dropouts in the communications at the same time to meet the performance requirement is introduced in Chap. 8.

2.2.2 An Adaptive Event-Triggered Communication Scheme

An adaptive event-triggered communication scheme (AETC) is provided in this book to reduce the number of the transmitted data, while preserving the desired control performance. Different from some existing communication schemes with a constant threshold [16, 68, 90], the key feature of AETC is that the threshold condition of AETC can be adaptively adjusted based on the latest available state-dependent information.

Suppose that the occurrence of a transmission depends on a predesigned condition rather than the passing of time, which determines when the next transmission should be taken place. Figure 2.6 depicts a general diagram of the proposed AETC for an NCS, where the adaptive EG has the logic function to determine whether the sampled data should be transmitted to the controller or not; when the sampled data is transmitted over the communication networks, the threshold of ETC scheme is calculated and stored at Storer 1, and the sampled data is also stored at Storer 2 for the next calculation of the state-dependent error. Compared with some existing ETC schemes [16, 67, 68], the main advantage of the proposed AETC is that the threshold condition can be dynamically adjusted to save more limited communication bandwidth, while ensuring the desired control performance. Based the above description, the next transmission instant determined by the above adaptive EG can be expressed as

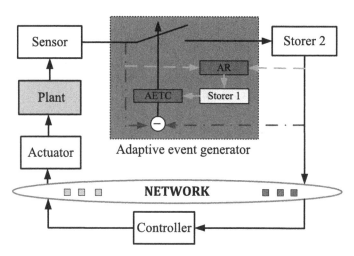

Fig. 2.6 A diagram of an adaptive event-triggered scheme for an NCS. *AR* adaptive rule; *AETC* adaptive event-triggered communication scheme

$$t_{k+1}h = t_kh + \min_{\ell \in \mathbb{N}}\{\ell h | e^T(i_kh)\Phi e(i_kh) > \sigma(t_kh)x^T(t_kh)\Phi x(t_kh)\} \quad (2.17)$$

where $\Phi > 0$ is a weighting matrix, and $i_kh = t_kh + \ell h$, $\ell \in \mathbb{N}$, h is the sampling period; t_k $(k = 0, 1, 2, \ldots)$ are some integers such that $\{t_0, t_1, t_2, \ldots\} \subset \{0, 1, 2, \ldots\}$; $e(i_kh)$ is the error between the current sampled data $x(i_kh)$ and the latest transmission data $x(t_kh)$, that is

$$e(i_kh) \triangleq x(i_kh) - x(t_kh), i_kh \in (t_k, t_{k+1}] \quad (2.18)$$

Moreover, $\sigma(t_kh)$ in (2.17) is determined by the following adaptive rule

$$\sigma(t_{k+1}h) = \max\{\underbrace{\sigma(t_kh)(1 - \frac{2\alpha}{\pi}\text{atan}[\beta(\|x(t_{k+1}h)\| - \|x(t_kh)\|)])}_{\lambda}, \sigma_m\} \quad (2.19)$$

where atan$(\cdot)$ is the invert tangent function, $0 < \alpha$ and $0 < \beta$ are given constants to adjust the output of atan$(\cdot)$, σ_m is the given lower bound of $\sigma(t_kh)$, $\sigma(0) = \sigma_m$.

From (2.17) and (2.19), one can see that the transmission events are dependent both on the error $e(i_kh)$ and the latest transmitted states $x(t_kh)$, and on the adjustable threshold $\sigma(t_kh)$.

Different from the ETC scheme (2.13), where the ETC thresholds σ_1 is a preselected constant. One can see that the parameter $\sigma(t_kh)$ in (2.17) can be adaptively adjusted based on (2.19), which is dependent on the current $x(t_{k+1}h)$, the latest $x(t_kh)$, α, β and σ_m simultaneously. Therefore, the proposed AETC (2.17) has the advantage in potentially reducing the number of transmitted packets than those based on (2.19) in [16, 67, 68] while ensuring the desired control performance.

2.2.3 Delay Distribution-Dependent Model of NCSs

Based on the communication schemes mentioned above, this section introduces a communication-dependent delay model of NCSs, which has the advantage in unifying the ETC scheme with the other part of the system to be controlled.

Consider a class of systems governed by

$$\dot{x}(t) = f(x(t), u(t), \omega(t)) \quad (2.20)$$

where $f(\cdot)$ is a function related with $x(t)$, $u(t)$ and $\omega(t)$; $x(t) \in \mathbb{R}^n$ and $u(t) \in \mathbb{R}^m$ are state vector and control input vector respectively; $\omega(t) \in \mathcal{L}_2[0, \infty)$ denotes the exogenous disturbance signal; the initial condition of the system (2.20) is given by $x(t_0) = x_0$. Throughout this book, it is assumed that the system (2.20) is controlled over a communication network. For showing the main idea of modeling method, a

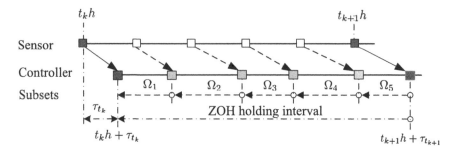

Fig. 2.7 Illustration of subsets of the ZOH. Reprinted from Ref.[69], Copyright 2013, with permission from Elsevier

networked state feedback controller is directly connected to the actuator through a ZOH. This leads to:

$$\begin{cases} \dot{x}(t) = f(x(t), u(t), \omega(t)) \\ u(t) = Kx(t_kh), t \in \Omega \end{cases} \tag{2.21}$$

where K is the state feedback controller gain; τ_{t_k} is the packet transmission delay; h is the sampling period, $\Omega =: [t_kh + \tau_{t_k}, t_{k+1}h + \tau_{t_{k+1}})$.

Divide the holding interval of the ZOH $t \in \Omega$ into sampling-interval-like subsets $\Omega_\ell = [i_kh + \tau_{i_k}, i_kh + h + \tau_{i_k+1})$ [16], i.e., $\Omega = \cup\Omega_\ell$, where $i_kh = t_kh + \ell h$, $\ell = 0, \ldots, t_{k+1} - t_k - 1$ means the sampling instants from the current transmitted sampling instant t_kh to the future transmitted sampling instant $t_{k+1}h$; if ℓ takes the value of $t_{k+1} - t_k - 1$, then $\tau_{i_k+1} = \tau_{t_{k+1}}$, otherwise, $\tau_{i_k} = \tau_{t_k}$. See Fig. 2.7 for an illustration. Define $\eta(t) \triangleq t - i_kh, t \in \Omega_\ell$. It is clear that $\eta(t)$ is a piecewise linear function satisfying

$$0 < \eta_1 \leq \eta(t) \leq h + \bar{\tau} = \eta_3, \ t \in \Omega_\ell \tag{2.22}$$

where $\eta_1 = inf_\ell\{\tau_{i_k}\}$, $\eta_3 = h + sup_\ell\{\tau_{i_k+1}\} = h + \bar{\tau}$; h and $\bar{\tau}$ are the sampling period and the maximum allowable upper communication delay bound, respectively. Then, the control of (2.21) is:

$$u(t) = K(x(t - \eta(t)) - e(i_kh)), \ t \in \Omega_\ell \tag{2.23}$$

Combining (2.21) and (2.23) leads to a sampled-state-error dependent closed-loop NCS model

$$\dot{x}(t) = f(x(t), K(x(t - \eta(t)) - e(i_kh)), \omega(t)), \ t \in \Omega_\ell \tag{2.24}$$

where the initial condition of the state $x(t)$ is: $x(t) = \phi(t), t \in [t_0 - \eta_3, t_0], \phi(t_0) = x_0$, and $\phi(t)$ is a continuous function on $[t_0 - \eta_3, t_0]$.

Combining ETC schemes, i.e., (2.13), (2.16), (2.17) and the system model (2.24), communication-dependent analysis and synthesis methods will be used in Chaps. 6–8 to consider the communication and control in a unified framework.

2.3 Main Mathematical Lemmas

In this section, some main lemmas are provided to derive the main results of this book. The proof of these lemmas can be seen in the literature, therefore, they are omitted in this book.

Lemma 2.1 (Schur complement). *For a symmetric matrix* $S = \begin{bmatrix} S_{11} & S_{12} \\ S_{21} & S_{22} \end{bmatrix}$, *where* S_{11} *and* $S_{22} \in \mathbb{R}^{n \times n}$. *Then the following conditions are equivalent:*

1. $S < 0$;
2. $S_{11} < 0$, $S_{22} - S_{12}^T S_{11}^{-1} S_{12} < 0$;
3. $S_{22} < 0$, $S_{11} - S_{12} S_{22}^{-1} S_{12}^T < 0$.

Lemma 2.2 (Jensen's inequality) [91]. *For any constant matrix* $W \in \mathbb{R}^{n \times n}$, $W = W^T \geq 0$, *scalar* $0 \leq \tau_1 \leq \tau_2$, *and vector-valued function* $\dot{x}:[-\tau_2, -\tau_1] \to \mathbb{R}^n$ *such that the following integration is well defined, it holds that*

$$- (\tau_2 - \tau_1) \int_{t-\tau_2}^{t-\tau_1} \dot{x}^T(s) W \dot{x}(s) ds \leq - \begin{bmatrix} x(t-\tau_1) \\ x(t-\tau_2) \end{bmatrix}^T \begin{bmatrix} W & * \\ -W & W \end{bmatrix} \begin{bmatrix} x(t-\tau_1) \\ x(t-\tau_2) \end{bmatrix} \tag{2.25}$$

Lemma 2.3 ([92, 93]). *For any constant matrix* $R \in \mathbb{R}^{n \times n}$, $R = R^T$, *matrix* $S \in \mathbb{R}^{n \times n}$, $\begin{bmatrix} R & * \\ S & R \end{bmatrix} \geq 0$, *scalars* $0 \leq \tau_1 \leq \tau(t) \leq \tau_2$ *and vector function* $\dot{x} : [-\tau_2, -\tau_1] \to \mathbb{R}^n$ *such that the following integration is well defined, it holds that*

$$- (\tau_2 - \tau_1) \int_{t-\tau_2}^{t-\tau_1} \dot{x}^T(s) R \dot{x}(s) ds \leq - \sum_{i=1}^{4} \Xi_4 \tag{2.26}$$

where

$$\Xi_1 = [x^T(t-\tau_1) - x^T(t-\tau(t))] R [x(t-\tau_1) - x(t-\tau(t))]$$
$$\Xi_2 = [x^T(t-\tau(t)) - x^T(t-\tau_2)] R [x(t-\tau(t)) - x(t-\tau_2)]$$
$$\Xi_3 = [x^T(t-\tau_1) - x^T(t-\tau(t))] S^T [x(t-\tau(t)) - x(t-\tau_2)]$$
$$\Xi_4 = [x^T(t-\tau(t)) - x^T(t-\tau_2)] S [x(t-\tau_1) - x(t-\tau(t))]$$

Lemma 2.4 ([94]). *Suppose* $\tau_m \leq \tau(t) \leq \tau_M$. *For any constant matrices* Ξ_1, Ξ_2 *and* Ω, *then*

$$(\tau(t) - \tau_m)\varXi_1 + (\tau_M - \tau(t))\varXi_2 + \varOmega < 0 \tag{2.27}$$

holds, if and only if

$$(\tau_M - \tau_m)\varXi_1 + \varOmega < 0 \tag{2.28}$$

$$(\tau_M - \tau_m)\varXi_2 + \varOmega < 0 \tag{2.29}$$

hold.

Lemma 2.5 ([85]). *For any constant matrix $Q \in \mathbb{R}^{n \times n}, Q > 0$, scalars $\tau_1 \leq \tau(t) \leq \tau_3$, and vector function $\dot{x} : [-\tau_3, -\tau_1] \to \mathbb{R}^n$ such that the following integration is well defined, then it holds that*

$$- (\tau_3 - \tau_1) \int_{t-\tau_3}^{t-\tau_1} \dot{x}^T(v) Q \dot{x}(v) dv \leq \zeta_1^T(t) \begin{bmatrix} -Q & Q & 0 \\ * & -2Q & Q \\ * & * & -Q \end{bmatrix} \zeta_1(t) \tag{2.30}$$

where $\zeta_1^T(t) = \left[x^T(t-\tau_1) \; x^T(t-\tau(t)) \; x^T(t-\tau_3) \right]$.

Lemma 2.6 ([95]). *Let A, L, E, and F be real matrices of appropriate dimensions, with F satisfying $\|F\| \leq 1$. Then we have the following inequalities:*

1. *For any scalar $\varepsilon > 0$,*

$$LFE + E^T F^T L^T \leq \varepsilon^{-1} L L^T + \varepsilon E^T E$$

2. *For any matrix $P > 0$ and any scalar $\varepsilon > 0$ such that $P - \varepsilon L L^T > 0$,*

$$(A + LFE)^T P^{-1} (A + LFE) \leq A^T (P - \varepsilon L L^T)^{-1} A + \varepsilon^{-1} E^T E$$

Lemma 2.7 ([93]). *For any constant matrices $R \in \mathbb{R}^{n \times n}$ and $U \in \mathbb{R}^{n \times n}$, $\mathbb{U} = \begin{bmatrix} R & U^T \\ U & R \end{bmatrix} \geq 0$, scalars $0 \leq \eta_1 \leq \eta(t) \leq \eta_2$, and vector function $\mathfrak{y} : [-\eta_2, -\eta_1] \to \mathbb{R}^n$, such that the following sum is well defined, then*

$$\sum_{j=k-\eta_2}^{k-\eta_1-1} \mathfrak{y}^T(j) R \mathfrak{y}(j) \geq (\eta_2 - \eta_1) \begin{bmatrix} x_2 - x_3 \\ x_3 - x_4 \end{bmatrix}^T \mathbb{U} \begin{bmatrix} x_2 - x_3 \\ x_3 - x_4 \end{bmatrix} \tag{2.31}$$

where $x_2 = \tilde{x}(k-\eta_1), x_3 = \tilde{x}(k-\eta(k)), x_4 = \tilde{x}(k-\eta_2), \mathfrak{y}(k) = \tilde{x}(k+1) - \tilde{x}(k)$.

Part II
Communication-Delay-Distribution Dependent Method for NCSs

Chapter 3
Delay Distribution-Dependent Control for Networked Linear Control Systems

IP-based network delays in networked control systems (NCSs) are inherently nonuniformly distributed and behave with multifractal nature. This chapter proposes a delay distribution-based stability analysis and synthesis approach for a linear system controlled over an IP-based communication network. To fully use the nonuniform distribution characteristics of IP-based network delays, a stochastic control model related with the characteristics of communication networks is established to describe the NCSs. Then, delay distribution-dependent NCS stability criteria are derived in the form of Linear Matrix Inequalities (LMIs). Also, the maximum allowable upper delay bound and controller feedback gain can be obtained simultaneously from the developed approach by solving a constrained convex optimization problem. Numerical examples showed that the results derived from the proposed method are less conservative than those derived from some existing methods [85].

This chapter is organized as follows. A delay distribution-based control strategy for NCSs is addressed in Sect. 3.1. Stability and stabilization of the studied system is studied in Sect. 3.2. Section 3.3 gives an example to show the effectiveness of the proposed method. Finally, Sect. 3.4 concludes the chapter.

3.1 A Delay Distribution-Based Control Strategy for NCSs

Consider a class of linear systems governed by

$$\dot{x}(t) = Ax(t) + Bu(t) \tag{3.1}$$

$$x(t) = \rho(t), \quad t \in [-\tau_3, 0] \tag{3.2}$$

where $x(t) \in \mathbb{R}^n$ and $u(t) \in \mathbb{R}^m$ are state vector and control input vector, respectively. A and B are constant matrices with appropriate dimensions. $\rho(t)$ is the given function which means the initial condition of the systems on the segment of $t \in [-\tau_3, 0]$.

© Springer-Verlag Berlin Heidelberg 2015
C. Peng et al., *Communication and Control for Networked Complex Systems*,
DOI 10.1007/978-3-662-46813-5_3

Based on the delay distribution-dependent modeling method described in Sect. 2.1, the following model for the closed-loop systems controlled over an IP-based network is obtained:

$$\begin{cases} \dot{x}(t) = Ax(t) + \delta(t)BKx(t - \tau_1(t)) + (1 - \delta(t))BKx(t - \tau_2(t)) \\ x(t) = \rho(t), \quad t \in [-\tau_3, 0] \end{cases} \tag{3.3}$$

To facilitate developments, Eq. (3.3) is rewritten as

$$\begin{aligned} \dot{x}(t) &= Ax(t) + \bar{\delta}BKx(t - \tau_1(t)) + (1 - \bar{\delta})BKx(t - \tau_2(t)) \\ &\quad + (\delta(t) - \bar{\delta})BK[x(t - \tau_1(t)) - x(t - \tau_2(t))] \\ &= \varphi(t) + (\delta(t) - \bar{\delta})\psi(t) \end{aligned} \tag{3.4}$$

with

$$\varphi(t) \triangleq Ax(t) + \bar{\delta}BKx(t - \tau_1(t)) + (1 - \bar{\delta})BKx(t - \tau_2(t))$$
$$\psi(t) \triangleq BKx(t - \tau_1(t)) - BKx(t - \tau_2(t))$$

In the following, some practically computable stability criteria are developed for the NCSs described by (3.3).

3.2 Stability and Stabilization Results

In this section, the stability and synthesis conditions are presented for closed-loop system (3.3). Assume that the NCSs network characteristics can be described by (2.2), the following theorem shows that the closed-loop system is asymptotically stable in the mean square if particular LMIs are feasible [85].

Theorem 3.1 *For given constants τ_i ($i = 1, 2, 3$) and matrix K, if there exist matrices $P > 0$, $Q_i > 0$, $R_i > 0$, $S_j > 0$, and U_j, ($i = 1, 2, 3, j = 1, 2$) with appropriate dimensions, such that the following LMIs hold for $j = 1, 2$, then system* (3.3) *is asymptotically stable in the mean square.*

$$\Pi = \begin{bmatrix} \Pi_{11} & * & * \\ \Pi_{21} & \Pi_{22} & * \\ \Pi_{31} & 0 & \Pi_{33} \end{bmatrix} < 0 \tag{3.5}$$

$$\begin{bmatrix} S_j & * \\ U_j & S_j \end{bmatrix} > 0 \tag{3.6}$$

where

$$\Pi_{11} = \begin{bmatrix} \Sigma_{11} & * & * & * & * & * \\ \Sigma_{21} & \Sigma_{22} & * & * & * & * \\ \Sigma_{31} & \Sigma_{32} & \Sigma_{33} & * & * & * \\ 0 & 0 & \Sigma_{43} & \Sigma_{44} & * & * \\ \Sigma_{51} & 0 & 0 & \Sigma_{54} & \Sigma_{55} & * \\ 0 & 0 & 0 & 0 & \Sigma_{65} & \Sigma_{66} \end{bmatrix},$$

$$\Pi_{21} = diag\{A, 0, \bar{\delta}BK, 0, (1-\bar{\delta})BK, 0\}Ones(6,5)\Xi,$$
$$\Pi_{31} = \bar{\delta}(1-\bar{\delta})diag\{0, 0, BK, 0, -BK, 0\}Ones(6,5)\Xi,$$
$$\Pi_{22} = diag\{-R_1, -R_2, -R_3, -S_1, -S_2\}, \quad \Pi_{33} = \bar{\delta}(1-\bar{\delta})\Pi_{22},$$
$$\Xi = diag\{\tau_1 R_1, \tau_2 R_2, \tau_3 R_3, (\tau_2 - \tau_1)S_1, (\tau_3 - \tau_2)S_2\},$$

and

$$\Sigma_{11} = PA + A^T P + Q_1 + Q_2 + Q_3 - R_1 - R_2 - R_3,$$
$$\Sigma_{12} = R_1, \quad \Sigma_{22} = -Q_1 - R_1 - S_1, \quad \Sigma_{31} = \bar{\delta}K^T B^T P + R_2,$$
$$\Sigma_{32} = S_1 - U_1, \quad \Sigma_{33} = -2R_2 - 2S_1 + U_1 + U_1^T, \quad \Sigma_{43} = U_1,$$
$$\Sigma_{43} = R_2 + S_1 - U_1, \quad \Sigma_{44} = -Q_2 - R_2 - S_1 - S_2,$$
$$\Sigma_{51} = (1-\bar{\delta})K^T B^T P + R_3, \quad \Sigma_{54} = S_2 - U_2,$$
$$\Sigma_{55} = -2R_3 - 2S_2 + U_2 + U_2^T, \quad \Sigma_{64} = U_2,$$
$$\Sigma_{65} = R_3 + S_2 - U_2, \quad \Sigma_{66} = -Q_3 - R_3 - S_2.$$

Ones(6, 5) is a 6-by-5 matrix of ones.

Proof 1 Construct a Lyapunov–Krasovskii functional candidate as

$$V(x_t) = \sum_{i=1}^{4} V_i(x_t) \tag{3.7}$$

where

$$V_1(x_t) = x^T(t)Px(t),$$
$$V_2(x_t) = \sum_{i=1}^{3} \int_{t-\tau_i}^{t} x^T(s)Q_i x(s)ds,$$
$$V_3(x_t) = \sum_{i=1}^{3} \int_{-\tau_i}^{0} \int_{t+s}^{t} \dot{x}^T(v)\tau_i R_i \dot{x}(v)dvds$$
$$V_4(x_t) = \sum_{i=1}^{2} \int_{-\tau_{i+1}}^{-\tau_i} \int_{t+s}^{t} \dot{x}^T(v)(\tau_{i+1} - \tau_i)S_i \dot{x}(v)dvds$$

Q_i, R_i, S_j ($i = 1, 2, 4$, $j = 1, 2$) > 0. From Eq. (2.4), it is clear that $\mathbb{E}\{\delta(t) - \bar{\delta}\} = 0$, $\mathbb{E}\{(\delta(t) - \bar{\delta})^2\} = \bar{\delta}(1 - \bar{\delta})$. Then, the mathematical expectation of the generator $\dot{V}(x_t)$ for the evolution of $V(x_t)$ along the solutions of system (3.3) is given by

$$\mathbb{E}\{\dot{V}_1(x_t)\} = 2\mathbb{E}\{x^T(t)P\dot{x}(t)\}$$
$$= 2\mathbb{E}\{x^T(t)P\varphi(t) + (\delta - \bar{\delta})x^T(t)P\psi(t)\}$$
$$= 2x^T(t)P\varphi(t) \tag{3.8}$$

$$\mathbb{E}\{\dot{V}_2(x_t)\} = \sum_{i=1}^{3}\{x^T(t)Q_i x(t) - x^T(t - \tau_i)Q_i x(t - \tau_i)\} \tag{3.9}$$

According to (3.3), we have

$$\mathbb{E}\{\dot{V}_3(x_t)\} = \mathbb{E}\{\sum_{i=1}^{3}\{\dot{x}^T(t)\tau_i^2 R_i \dot{x}(t) - \int_{t-\tau_i}^{t}\dot{x}^T(v)\tau_i R_i \dot{x}(v)dv\}\} \tag{3.10}$$

where

$$\mathbb{E}\{\sum_{i=1}^{3}\dot{x}^T(t)\tau_i^2 R_i \dot{x}(t)\} = \mathbb{E}\{\sum_{i=1}^{3}[\varphi(t) + (\delta - \bar{\delta})\psi(t)]^T \tau_i^2 R_i [\varphi(t) + (\delta - \bar{\delta})\psi(t)]\}$$

$$= \sum_{i=1}^{3}\varphi^T(t)\tau_i^2 R_i \varphi(t) + \bar{\delta}(1 - \bar{\delta})\sum_{i=1}^{3}\psi^T(t)\tau_i^2 R_i \psi(t)$$

Applying Lemmas 2.1 and 2.5 in Chap. 2, when $R_i > 0$, we have

$$-\int_{t-\tau_1}^{t}\dot{x}^T(v)\tau_1 R_1 \dot{x}(v)dv \leq \zeta_1^T(t)\begin{bmatrix} -R_1 & R_1 \\ * & -R_1 \end{bmatrix}\zeta_1(t) \tag{3.11}$$

$$-\int_{t-\tau_2}^{t}\dot{x}^T(v)\tau_2 R_2 \dot{x}(v)dv \leq \zeta_2^T(t)\begin{bmatrix} -R_2 & R_2 & 0 \\ * & -2R_2 & R_2 \\ * & * & -R_2 \end{bmatrix}\zeta_2(t) \tag{3.12}$$

$$-\int_{t-\tau_3}^{t}\dot{x}^T(v)\tau_3 R_3 \dot{x}(v)dv \leq \zeta_3^T(t)\begin{bmatrix} -R_3 & R_3 & 0 \\ * & -2R_3 & R_3 \\ * & * & -R_3 \end{bmatrix}\zeta_3(t) \tag{3.13}$$

where $\zeta_1(t) = \begin{bmatrix} x(t) \\ x(t - \tau_1) \end{bmatrix}$, $\zeta_2(t) = \begin{bmatrix} x(t) \\ x(t - \tau_1(t)) \\ x(t - \tau_2) \end{bmatrix}$, $\zeta_3(t) = \begin{bmatrix} x(t) \\ x(t - \tau_2(t)) \\ x(t - \tau_3) \end{bmatrix}$.

According to (3.3), we have

$$\mathbb{E}\{\dot{V}_4(x_t)\} = \mathbb{E}\{\sum_{i=1}^{2}\{\dot{x}^T(t)(\tau_{i+1} - \tau_i)^2 S_i \dot{x}(t) - \int_{t-\tau_{i+1}}^{t-\tau_i} \dot{x}^T(v)(\tau_{i+1} - \tau_i)S_i\dot{x}(v)dv\}\}$$

(3.14)

where

$$\mathbb{E}\{\sum_{i=1}^{2}\dot{x}^T(t)(\tau_{i+1} - \tau_i)^2 S_i \dot{x}(t)\}$$

$$= \mathbb{E}\{\sum_{i=1}^{2}[\varphi(t) + (\delta - \bar{\delta})\psi(t)]^T(\tau_{i+1} - \tau_i)^2 S_i[\varphi(t) + (\delta - \bar{\delta})\psi(t)]\}$$

$$= \sum_{i=1}^{2}\varphi^T(t)(\tau_{i+1} - \tau_i)^2 S_i\varphi(t) + \bar{\delta}(1 - \bar{\delta})\sum_{i=1}^{2}\psi^T(t)(\tau_{i+1} - \tau_i)^2 S_i\psi(t) \quad (3.15)$$

Applying Lemma 2.3 in Chap. 2, we have

$$-\int_{t-\tau_2}^{t-\tau_1} \dot{x}^T(v)(\tau_2 - \tau_1)S_1\dot{x}(v)dv \leq -\zeta_4^T(t)\begin{bmatrix} S_1 & * \\ U_1 & S_1 \end{bmatrix}\zeta_4(t) \quad (3.16)$$

$$-\int_{t-\tau_3}^{t-\tau_2} \dot{x}^T(v)(\tau_3 - \tau_2)S_2\dot{x}(v)dv \leq -\zeta_5^T(t)\begin{bmatrix} S_2 & * \\ U_2 & S_2 \end{bmatrix}\zeta_5(t) \quad (3.17)$$

where $\zeta_4(t) = \begin{bmatrix} x(t-\tau_1) - x(t-\tau_1(t)) \\ x(t-\tau_1(t)) - x(t-\tau_2) \end{bmatrix}$, $\zeta_5(t) = \begin{bmatrix} x(t-\tau_2) - x(t-\tau_2(t)) \\ x(t-\tau_2(t)) - x(t-\tau_3) \end{bmatrix}$.

Then, considering (3.8)–(3.17) together, we have

$$\mathbb{E}\{\dot{V}(x_t)\} \leq \xi^T(t)\Pi_{11}\xi(t) + \sum_{i=1}^{3}\varphi^T(t)\tau_i^2 R_i\varphi(t) + \bar{\delta}(1 - \bar{\delta})\sum_{i=1}^{3}\psi^T(t)\tau_i^2 R_i\psi(t)$$

$$+ \sum_{i=1}^{2}\varphi^T(t)(\tau_{i+1} - \tau_i)^2 S_i\varphi(t) + \bar{\delta}(1 - \bar{\delta})\sum_{i=1}^{2}\psi^T(t)(\tau_{i+1} - \tau_i)^2 S_i\psi(t)$$

$$= \xi^T(t)[\Pi_{11} - \Pi_{21}^T\Pi_{22}^{-1}\Pi_{21} - \Pi_{31}^T\Pi_{33}^{-1}\Pi_{31}]\xi(t) \quad (3.18)$$

where $\xi^T(t) = [x^T(t), x^T(t-\tau_1), x^T(t-\tau_1(t)), x^T(t-\tau_2), x^T(t-\tau_2(t)), x^T(t-\tau_3)]$, $\Pi_{ij}(i, j = 1, 2, 3)$ is defined in Theorem 3.1.

By Schur complement, (3.5) implies that

$$\Pi_{11} - \Pi_{12}\Pi_{22}^{-1}\Pi_{12}^T - \Pi_{13}\Pi_{33}^{-1}\Pi_{13}^T < 0 \quad (3.19)$$

Then, considering (3.8)–(3.19) together, we have $\mathbb{E}\{\dot{V}(x_t)\} < 0$. It can be concluded from Lyapunov stability theory that the dynamics of system (3.3) is asymptotically stable in the mean square. This completes the proof. $\qquad\square$

Based on Theorem 3.1, the stabilization criterion is summarized as follows:

Theorem 3.2 *For given constants $\tau_i (i = 1, 2, 3)$, if there exist matrices $X > 0$, $\tilde{Q}_i > 0$, $\tilde{R}_i > 0$ and $\tilde{S}_j > 0$, and $\tilde{U}_j$, $(i = 1, 2, 3, j = 1, 2)$, with appropriate dimensions, such that the following matrix inequalities hold for $j = 1, 2$, then system (3.3) is asymptotically stable in the mean square with controller feedback gain $K = YX^{-1}$.*

$$\begin{bmatrix} \tilde{\Pi}_{11} & * & * \\ \tilde{\Pi}_{21} & \tilde{\Pi}_{22} & * \\ \tilde{\Pi}_{31} & 0 & \tilde{\Pi}_{33} \end{bmatrix} < 0 \qquad (3.20)$$

$$\begin{bmatrix} \tilde{S}_j & * \\ \tilde{U}_j & \tilde{S}_j \end{bmatrix} > 0 \qquad (3.21)$$

where

$$\tilde{\Pi}_{11} = \begin{bmatrix} \tilde{\Sigma}_{11} & * & * & * & * & * \\ \tilde{\Sigma}_{21} & \tilde{\Sigma}_{22} & * & * & * & * \\ \tilde{\Sigma}_{31} & \tilde{\Sigma}_{32} & \tilde{\Sigma}_{33} & * & * & * \\ 0 & 0 & \Sigma_{43} & \tilde{\Sigma}_{44} & * & * \\ \tilde{\Sigma}_{51} & 0 & 0 & \Sigma_{54} & \tilde{\Sigma}_{55} & * \\ 0 & 0 & 0 & 0 & \tilde{\Sigma}_{65} & \tilde{\Sigma}_{66} \end{bmatrix},$$

$$\tilde{\Pi}_{21} = diag\{AX, 0, \bar{\delta}BY, 0, (1 - \bar{\delta})BK, 0\}Ones(6, 5)\tilde{\Xi},$$

$$\tilde{\Pi}_{31} = \bar{\delta}(1 - \bar{\delta})diag\{0, 0, BY, 0, -BY, 0\}Ones(6, 5)\tilde{\Xi},$$

$$\tilde{\Pi}_{22} = diag\{-X\tilde{R}_1^{-1}X, -X\tilde{R}_2^{-1}X, -X\tilde{R}_3^{-1}X, -X\tilde{S}_1^{-1}X, -X\tilde{S}_2^{-1}X\},$$

$$\tilde{\Pi}_{33} = \bar{\delta}(1 - \bar{\delta})\tilde{\Pi}_{22},$$

$$\tilde{\Xi} = diag\{\tau_1 I, \tau_2, \tau_3 I, (\tau_2 - \tau_1)I, (\tau_3 - \tau_2)I\},$$

and

$$\tilde{\Sigma}_{11} = AX + XA^T + \tilde{Q}_1 + \tilde{Q}_2 + \tilde{Q}_3 - \tilde{R}_1 - \tilde{R}_2 - \tilde{R}_3,$$

$$\tilde{\Sigma}_{12} = \tilde{R}_1, \ \tilde{\Sigma}_{22} = -\tilde{Q}_1 - \tilde{R}_1 - \tilde{S}_1, \ \tilde{\Sigma}_{31} = \bar{\delta}Y^T B^T + \tilde{R}_2,$$

$$\tilde{\Sigma}_{32} = \tilde{S}_1 - \tilde{U}_1, \ \tilde{\Sigma}_{33} = -2\tilde{R}_2 - 2\tilde{S}_1 + \tilde{U}_1 + \tilde{U}_1^T, \ \tilde{\Sigma}_{43} = \tilde{U}_1,$$

$$\tilde{\Sigma}_{43} = \tilde{R}_2 + \tilde{S}_1 - \tilde{U}_1, \ \tilde{\Sigma}_{44} = -\tilde{Q}_2 - \tilde{R}_2 - \tilde{S}_1 - \tilde{S}_2,$$

$$\tilde{\Sigma}_{51} = (1 - \bar{\delta})Y^T B^T + \tilde{R}_3, \ \tilde{\Sigma}_{54} = \tilde{S}_2 - \tilde{U}_2,$$

$$\tilde{\Sigma}_{55} = -2\tilde{R}_3 - 2\tilde{S}_2 + \tilde{U}_2 + \tilde{U}_2^T, \ \tilde{\Sigma}_{64} = \tilde{U}_2,$$

$$\tilde{\Sigma}_{65} = \tilde{R}_3 + \tilde{S}_2 - \tilde{U}_2, \ \tilde{\Sigma}_{66} = -\tilde{Q}_3 - \tilde{R}_3 - \tilde{S}_2.$$

Proof 2 Pre and postmultiply both sides of (3.5) and (3.6) with $diag(X_{1\times6}, I_{1\times10})$, $diag(X_{1\times2})$, and their transposes. From (3.5), If $\Pi < 0$, it follows that Σ_{11} must be negative definite, which leads to P being nonsigular, then defining $Y = KX$, $X = P^{-1}$, $\tilde{Q}_i = XQ_iX$, $\tilde{R}_i = XR_iX$, $(i = 1, 2, 3)$, $\tilde{S}_j = XS_jX$, $\tilde{U}_j = XU_jX$, $(j = 1, 2)$, and applying Schur complement, we have (3.20). This completes the proof.   □

Notice that the matrix inequality (3.20) is nonconvex inequality because of the items $X\tilde{R}_1^{-1}X$ and $-X\tilde{R}_2^{-1}X$, etc., in $\tilde{\Pi}_{22}$, and it cannot be directly solved by MatLab LMIs toolbox. However, instead of the original nonconvex minimization problem, the nonlinear minimization based on cone complementary linearization algorithm can be adopted to solve this nonconvex problem [15, 96]. Although this gives only a suboptimal solution to the original problem (3.20), it is much easier to solve than the original nonconvex minimization problem.

3.3 An Example

This section aims to demonstrate the effectiveness of the proposed NCS control strategy and derived theorems using numerical examples. For comparisons with existing methods [12, 15, 97–100], we choose the NCSs governed by (3.1) and (3.2) as in these references. The system settings are

$$A = \begin{bmatrix} 0 & 1 \\ 0 & -0.1 \end{bmatrix}, B = \begin{bmatrix} 0 \\ 0.1 \end{bmatrix}, K = \begin{bmatrix} -3.75 & -11.5 \end{bmatrix} \quad (3.22)$$

When $\bar{\delta}$ equals to 0 or 1, Theorem 3.1 is reduced to the special case without the statistical distribution of network-induced delay $\tau(t)$. For comparisons with this case, using the existing methods in [12, 15, 97–100] and Theorem 3.1 in this chapter, the maximum allowable delay bounds (MADBs) were calculated for NCSs stability based on methods mentioned above. Tables 3.1 and 3.2 list the results with or without considering lower delay bounds, respectively.

One can see from Tables 3.1 and 3.2 that the results obtained in this chapter are less conservative than those obtained in [12, 15, 97–99].

When considering the stochastic distribution of the network-induced delay and applying Theorem 3.1, Table 3.3 lists the obtained maximum allowable τ_3.

Table 3.1 MADBs—for Theorem 3.1 with $\bar{\delta} = 0$ or 1, and $\tau_1 = 0$

Methods	[97]	[98]	[12]	[99]	Theorem 3.1
MADBs	0.0538	0.7805	0.8695	0.9410	1.0432

Table 3.2 MADBs—for Theorem 3.1 with $\bar{\delta} = 1$

Lower delay bound τ_1	$\tau_1 = 0$	$\tau_1 = 0.01$	$\tau_1 = 0.05$	$\tau_1 = 0.10$	$\tau_1 = 0.15$
Jiang [99], Peng [15],	0.9410	0.9421	0.9475	0.9520	0.9586
Theorem 3.1	1.0432	1.0433	1.0439	1.0448	1.0457

Table 3.3 The maximum allowable τ_3 under different settings of τ_2 and $\bar{\delta}$.

$\bar{\delta}(\tau_1 = 0)$	$\to 0$	0.4	0.6	0.8	$\to 1$
$\tau_2 = 0.02$	1.0435	1.3875	1.9382	3.3457	∞
$\tau_2 = 0.03$	1.0436	1.3872	1.9336	3.3401	∞
$\tau_2 = 0.04$	1.0438	1.3869	1.9289	3.3344	∞

From the above simulation, one can see that the more network characteristics are considered, the less conservative results can be obtained for NCSs stability. So, codesign of NCSs network and control is an effective scheme to improve the control performance of the integrated NCSs.

3.4 Conclusion

Network communication delays in IP-based NCSs are nonuniformly distributed and behave with the multifractal nature. The state feedback control for NCSs over an IP-based communication network has been investigated based on a Lyapunov–Krasovskii functional method and the Itô-differential rule. A delay distribution-based stochastic control coupled with QoC and QoP is developed. The criteria for stability and stabilization have been derived by introducing a tighter bounding technology and exploiting the information concerning interval distribution of the network-induced delay. Simulation results have demonstrated the feasibility and effectiveness of the proposed approach.

Chapter 4
Delay Distribution-Dependent Control for Networked Takagi–Sugeno Fuzzy Systems

In this chapter, parallel distributed compensation (PDC) fuzzy rules are investigated for Takagi–Sugeno (T-S) fuzzy systems under network environments subject to asynchronous grades of membership. First, a method is proposed to reconstruct the grades of membership at the controller, which guarantees that grades of membership at the fuzzy plant and at the fuzzy controller have the same timescale. Therefore, the problem of leading/lagging utilization of the latest information in the premises of PDC fuzzy rules in some existing ones is avoided, and the conservativeness induced by the asynchronous premise is reduced [51]. Second, the controller design specifically takes into account probabilistic interval distribution of the communication delay. If the probability distribution of communication delay is known or specified in a design process, based on the reconstructed grades of membership of the controller, sufficient stability conditions for networked T-S fuzzy systems are derived based on the Lyapunov theory. Following this, a stabilizing controller design method is developed. It shows that, the solvability of the design depends both on the upper and lower bounds of the delay and on its probability distribution [43].

The rest of this chapter is organized as follows. Section 4.1 presents an asynchronous premise reconstructed T-S fuzzy system model having a logic ZOH and incorporating the specific characteristics of communication delay. This random delay has a known or specified probability distribution. Section 4.2 analyzes the robust stability of the networked T-S fuzzy system modeled in Sect. 4.1, and develops a PDC controller design method subject to asynchronous grades of membership. A numerical example is given in Sect. 4.3. Finally, Sect. 4.4 concludes the chapter.

4.1 System and Problem Descriptions

Consider a T-S fuzzy system [101]. The ith rule of the system is expressed in the following IF-THEN form

© Springer-Verlag Berlin Heidelberg 2015
C. Peng et al., *Communication and Control for Networked Complex Systems*,
DOI 10.1007/978-3-662-46813-5_4

$$R^i : \text{If } \theta_1(t) \text{ is } W_1^i \text{ and}, \ldots, \text{ and } \theta_g(t) \text{ is } W_g^i$$
$$\text{Then } \dot{x}(t) = \bar{A}_i x(t) + \bar{B}_i u(t) \tag{4.1}$$

where $i = 1, 2, \ldots, r$, r is the number of IF-THEN rules; $x(t) \in \mathbb{R}^n$ and $u(t) \in \mathbb{R}^m$ are state vector and input vector, respectively; W_j^i $(i = 1, 2, \ldots, r; j = 1, 2, \ldots, g)$ are fuzzy sets; and $\theta_j(t)$ $(j = 1, 2, \ldots, g)$ represent premise variables. The initial condition of the system (4.1) is given by $x(t_0) = x_0$. Denoting $\theta(t) = [\theta_1(t), \ldots, \theta_g(t)]^T$, assume that $\theta(t)$ is either given or a function of $x(t)$, and does not depend on $u(t)$. $\bar{A}_i = A_i + \Delta A_i(t)$ and $\bar{B}_i = B_i + \Delta B_i(t)$; A_i and B_i $(i = 1, 2, \ldots, r)$ are constant matrices with compatible dimensions; $\Delta A_i(t)$ and $\Delta B_i(t)$ are time-varying matrices with appropriate dimensions, and are defined as

$$[\Delta A_i(t), \Delta B_i(t)] = H_i F_i(t)[E_{ai}, E_{bi}], \quad i = 1, 2, \ldots, r \tag{4.2}$$

where H_i and E_{ai}, E_{bi} are known constant real matrices with appropriate dimensions; $F_i(t)$ is an unknown real time-varying matrix with Lebesgue measurable elements bounded by

$$F_i^T(t) F_i(t) \leq I \tag{4.3}$$

By using a center-average defuzzifier, product inference, and a singleton fuzzifier, the global dynamics of T-S fuzzy system (4.1) can be inferred as

$$\dot{x}(t) = \sum_{i=1}^{r} \mu_i(\theta(t))[\bar{A}_i x(t) + \bar{B}_i u(t)] \tag{4.4}$$

where

$$\mu_i(\theta(t)) = \frac{h_i(\theta(t))}{\sum_{i=1}^{r} h_i(\theta(t))}, \quad h_i(\theta(t)) = \Pi_{j=1}^{g} W_j^i(\theta_j(t)) \tag{4.5}$$

and $W_j^i(\theta_j(t))$ is the membership value of $\theta_j(t)$ in W_j^i. It is seen from Eq. (4.5) that $\forall i \in \{1, 2, \ldots, r\}$, $\mu_i(\theta(t))$ has the following properties

$$\mu_i(\theta(t)) \geq 0, \quad \sum_{i=1}^{r} \mu_i(\theta(t)) = 1. \tag{4.6}$$

Throughout this chapter, assume that the system (4.1) is controlled through an IP-based network and the system state is available for feedback [43].

4.1.1 Networked T-S Control with Logic ZOH

Under the proposed framework, a general NCS model is shown in Fig. 4.1. Some techniques used and assumptions made are:

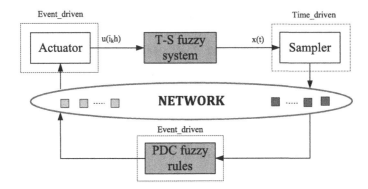

Fig. 4.1 A general diagram of networked T-S fuzzy control systems

- Each data packet is time stamped.
- The sampler is time-driven, i.e., the system state is sampled periodically at regular intervals.
- The controller is event-driven. It calculates the control signal as soon as it has received a packet from its uplink. Upon the completion of the calculation, it transmits a control packet to its downlink.
- The actuator is event-driven with a logic ZOH. The function of a logic ZOH is to select and to store the latest control signal based on the time stamps of received packets.

Notice that all the above are independent of the network-induced delay and packet dropout, this enables that all kinds of network-related uncertainties can be incorporated in an overall time delay (OTD). The maximum allowable OTD to maintain the system stability, i.e., the upper bound of OTD, is denoted τ_2 in this chapter. A design that leads to a higher τ_2 value is a less conservative design. This is achieved in this chapter mainly by fully utilizing the information about the delay, i.e., its probabilistic interval distribution.

In the following, a networked T-S fuzzy model-based controller will be designed via parallel distributed compensation (PDC) to stabilize T-S fuzzy system (4.1).

The rule for the ith controller is

$$R^i : \text{If} \quad \theta_1(t_k h) \text{ is } W_1^i \text{ and } \ldots \text{ and } \theta_g(t_k h) \text{ is } W_g^i$$
$$\text{Then } u(t^+) = K_j x(t_k h), \quad t \in \{t_k h + \tau_{t_k}, \, k = 1, 2, \ldots\} \tag{4.7}$$

where $u(t^+) = \lim_{\hat{t} \to t+0} u(\hat{t})$, h is the sampling period, t_k ($k = 1, 2, 3, \ldots$) are some integers such that $\{t_1, t_2, t_3, \ldots\} \subset \{0, 1, 2, 3, \ldots\}$, K_j ($j = 1, 2, \ldots, r$) are controller gains to be determined, τ_{t_k} is the network-induced delay. Here, it is assumed that the control computation and other overhead delays are included in τ_{t_k} [15].

Define the input delay $\tau(t) = t - t_k h$ for $t \in [t_k h + \tau_{t_k}, t_{k+1} h + \tau_{t_{k+1}})$, then

$$\tau_0 \leq \tau_{t_k} \leq \tau(t) < (t_{k+1} - t_k)h + \tau_{t_{k+1}} \leq \tau_2 \tag{4.8}$$

where $\tau_0 = \inf_k\{\tau_{t_k}\}$, $\tau_2 = \sup_k\left[(t_{k+1} - t_k)h + \tau_{t_{k+1}}\right]$. Therefore, the input delay in Logic ZOH can be represented by $u(t^+) = K_j x(t - \tau(t))$ in Eq. (4.7).

Because of the time stamp, the Logic ZOH guarantees that $t_{k+1} > t_k$. Moreover, the data dropout and disorder packets in the communication have been included in Eq. (4.7).

Based on the above analysis, the controller rule for the defuzzified output of (4.7) is designed as

$$u(t) = \sum_{j=1}^{r} \mu_j(\theta(t_k h)) K_j x(t - \tau(t)), \quad t \in \{t_k h + \tau_{t_k}, \ k = 1, 2, \ldots\} \quad (4.9)$$

Notice that the premise variables $\mu_j(\theta(t_k h))$ of the controller (4.9) are different from $\mu_j(\theta(t))$ in (4.1). That is, the premises of the plant and the networked PDC control rules are asynchronous. In this case, the obtained controller gains are identical, which shows that a PDC controller does not offer any advantages over a linear controller for the studied system controlled over a communication network [50, 51]. In the following, a method will be introduced to reconstruct $\mu_j(\theta(t_k h))$ in (4.9) to have the same time scales as $\mu_j(\theta(t))$ in (4.1) to obtain nonlinear PDC control rules.

4.1.2 Networked PDC Fuzzy Control with Asynchronous Premise Constraints

In this section, a method is presented to reconstruct the grades of the membership at the controller to avoid the problem of asynchronous premises mentioned above.

For notational simplicity, μ_j and μ_j^k are used to represent $\mu_j(\theta(t))$ and $\mu_j(\theta(t_k h))$ in the following description. Inspired by [51], the following asynchronous constraints on the membership functions are also used.

$$\begin{cases} \mu_j^k = \rho_j \mu_j \\ |\mu_j^k - \mu_j| \leq \Delta_j \end{cases} \quad (4.10)$$

where ρ_j, Δ_j $(j = 1, 2, \ldots, r)$ are some positive constants.

Based on the asynchronous constraints (4.10), it is clear that

$$v_1^j = 1 - \frac{\Delta_j}{\mu_j} \leq \rho_j \leq 1 + \frac{\Delta_j}{\mu_j} = v_2^j \quad (4.11)$$

where v_1^j and v_2^j denote the minimum and maximum values of ρ_j during the operation, which exhibits that

$$\frac{v_1^i}{v_2^j} = \frac{\min\{\rho_i\}}{\max\{\rho_j\}} \leq \min\{\frac{\rho_i}{\rho_j}\} \leq \frac{\rho_i}{\rho_j} \leq \max\{\frac{\rho_i}{\rho_j}\} \leq \frac{\max\{\rho_i\}}{\min\{\rho_j\}} = \frac{v_2^i}{v_1^j} \qquad (4.12)$$

Setting $v_1 = \min\{v_1^i\}$ and $v_2 = \max\{v_2^i\}$ $(i = 1, 2, \ldots, r)$, (4.12) yields that

$$\lambda_1 = \frac{v_1}{v_2} \leq \frac{\rho_i}{\rho_j} \leq \frac{v_2}{v_1} = \lambda_2 \qquad (4.13)$$

From (4.10), the controller (4.9) can be described as

$$u(t) = \sum_{j=1}^{r} \rho_j \mu_j K_j x(t - \tau(t)), \quad t \in \{t_k h + \tau_{t_k}, k = 1, 2, \ldots\} \qquad (4.14)$$

From the definition of λ_1 and λ_2 in (4.13), one can see that λ_2 equals to $\frac{1}{\lambda_1}$. As a special case, if $\lambda_1 = \lambda_2$, we have $\rho_i = \rho_j = \rho$ and $\mu_i^k = \rho\mu_i$ for $i = 1, 2, \ldots, r$. Based on the fact that $\sum_{i=1}^{r} \mu_i^k = \sum_{i=1}^{r} \mu_i = 1$, we have

$$1 = \sum_{i=1}^{r} \mu_i^k = \sum_{i=1}^{r} \rho\mu_i = \rho \sum_{i=1}^{r} \mu_i = \rho \qquad (4.15)$$

It follows that $\mu_i^k = \rho\mu_i = \mu_i$, which implies that the fuzzy controller always shares the same membership functions with the fuzzy model (4.1). In other words, when $\lambda_1 = \lambda_2 = 1$, the networked controller (4.9) is simplified as a point-to-point connected PDC controller.

Based on Eqs. (4.4) and (4.14), the continuity of the process in Fig. 4.1, together with the sampler and the logic ZOH, for $t \in [t_k h + \tau_{t_k}, t_{k+1} h + \tau_{t_{k+1}})$, the dynamics of the T-S fuzzy system with input delay is

$$\dot{x}(t) = \sum_{i=1}^{r} \sum_{j=1}^{r} \rho_j \mu_i \mu_j [\bar{A}_i x(t) + \bar{B}_i K_j x(t - \tau(t))] \qquad (4.16)$$

4.1.3 Delay Distribution-Dependent T-S Model

It is known that the overall delay $\tau(t)$ in Eq. (4.8) due to nonideal network is an interval time-varying delay and has its lower and upper bounds. From Chap. 2, it is known that the communication delays in IP-based NCSs have nonuniform distribution [82] and a multifractal nature [85, 86]. Therefore, to fully use the information of the characteristics of the network, a delay distribution-dependent feedback control law as described in Chap. 2.1 is adopted. This replaces the general form of the control law given in (4.9):

$$u(t) = \sum_{j=1}^{r} \rho_j \mu_j [\delta(t) K_j x(t - \tau_1(t)) + (1 - \delta(t)) K_j x(t - \tau_2(t))] \qquad (4.17)$$

where $t \in [t_k h + \tau_{t_k}, t_{k+1} h + \tau_{t_{k+1}})$, τ_1 is the medium value of the delay and assumed to be known; and

$$\begin{cases} \tau_1(t) = \delta(t)\tau(t), \text{ and } \delta(t) = 1 \text{ if } \tau(t) \in [\tau_0, \tau_1) \\ \tau_2(t) = (1 - \delta(t))\tau(t), \text{ and } \delta(t) = 0 \text{ if } \tau(t) \in [\tau_1, \tau_2) \end{cases} \qquad (4.18)$$

The $\tau_1(t)$ and $\tau_2(t)$ in (4.17) are important in the analysis of the networked closed-loop T-S fuzzy system. $\tau_1(t)$ means that the distribution of $\tau(t)$ is within a lower end range of $[\tau_0, \tau_1)$; while $\tau_2(t)$ means that the distribution of $\tau(t)$ is within a higher end range of $[\tau_1, \tau_2)$. It is also assumed that $\delta(t)$ in (4.17) is a bernoulli distributed sequence (BDS) as given in (2.4). Different from the work [102], where BDS is employed to deal with the systems with missing measurements, this paper utilizes BDS to describe the probability distribution of NCS communication delay.

Substituting (4.17) in (4.4) leads to the following closed-loop fuzzy system:

$$\begin{cases} \dot{x}(t) = \sum_{i=1}^{r} \mu_i \sum_{j=1}^{r} \rho_j \mu_j \{\bar{A}_i x(t) + \bar{B}_i [\delta(t) K_j x(t - \tau_1(t)) \\ \quad + (1 - \delta(t)) K_j x(t - \tau_2(t))]\}, \quad t \in [t_k h + \tau_{t_k}, t_{k+1} h + \tau_{t_{k+1}}) \end{cases} \qquad (4.19)$$

To facilitate further development, Eq. (4.19) can be rewritten as

$$\dot{x}(t) = \varphi(t) + (\delta(t) - \bar{\delta})\psi(t), \quad t \in [t_k h + \tau_{t_k}, t_{k+1} h + \tau_{t_{k+1}}) \qquad (4.20)$$

where

$$\begin{cases} \varphi(t) \triangleq \mathscr{A}^i x(t) + \bar{\delta} \mathscr{B}^i_j x(t - \tau_1(t)) + (1 - \bar{\delta}) \mathscr{B}^i_j x(t - \tau_2(t)) \\ \psi(t) \triangleq \mathscr{B}^i_j x(t - \tau_1(t)) - \mathscr{B}^i_j x(t - \tau_2(t)) \\ \mathscr{A}^i = \sum_{i=1}^{r} \mu_i \bar{A}_i, \quad \mathscr{B}^i_j = \sum_{i=1}^{r} \sum_{j=1}^{r} \rho_j \mu_i \mu_j \bar{B}_i K_j \end{cases}$$

4.2 Robust Stability Analysis and Controller Design

This section aims to develop an innovative approach for robust stability analysis of fuzzy system (4.1). For stability analysis, it is assumed that the feedback gain matrices K_j have been well designed.

4.2.1 Stability Analysis

First, consider a simple case without system parameter uncertainties $\Delta A_i(t)$ and $\Delta B_i(t)$. The stability condition is given below:

Theorem 4.1 *For given scalars λ_1 and λ_2, $\tau_0 \leq \tau_1 \leq \tau_2$ and $\bar{\delta}$, the system (4.20) is asymptotically stable in the mean square, if there exist matrices $P > 0$, $Q_i > 0$, $R_i > 0$ S_j, $(i = 1, 2, 3, j = 1, 2)$ with appropriate dimensions, such that the following LMIs hold for $l, i, j = 1, \ldots, r$, $1 \leq i < j \leq r$, and $\lambda = 0, 1, v = 1, 2$:*

$$\Sigma_{ll}(\lambda) < 0 \tag{4.21}$$

$$\Sigma_{ij}(\lambda) + \Sigma_{ji}^{v}(\lambda) < 0 \tag{4.22}$$

where

$$\Sigma_{ij}(\lambda) = \begin{bmatrix} F_{11}^{ij}(\lambda) & * & * \\ F_{21}^{ij} & F_{22} & * \\ F_{31}^{ij} & 0 & F_{33} \end{bmatrix}$$

$$\Sigma_{ij}^{v}(\lambda) = \begin{bmatrix} \lambda_v F_{11}^{ij}(\lambda) & * & * \\ F_{21}^{ij} & \frac{F_{22}}{\lambda_2} & * \\ F_{31}^{ij} & 0 & \frac{F_{33}}{\lambda_2} \end{bmatrix}$$

and

$$F_{11}^{ij}(\lambda) = \begin{bmatrix} \Xi_{11} & * & * & * & * & * \\ \Xi_{21} & \Xi_{22} & * & * & * & * \\ \Xi_{31} & \Xi_{32}(\lambda) & \Xi_{33}(\lambda) & * & * & * \\ 0 & \Xi_{42}(\lambda) & \Xi_{43}(\lambda) & \Xi_{44}(\lambda) & * & * \\ \Xi_{51}(\lambda) & 0 & 0 & \Xi_{54}(\lambda) & \Xi_{55}(\lambda) & * \\ 0 & 0 & 0 & \Xi_{64}(\lambda) & \Xi_{65}(\lambda) & \Xi_{66} \end{bmatrix}$$

$$F_{21}^{ij} = col\{R_1 \chi_1^{ij}, R_2 \chi_1^{ij}, R_3 \chi_1^{ij}, S_1 \chi_1^{ij}, S_2 \chi_1^{ij}\}$$

$$F_{31}^{ij} = col\{R_1 \chi_2^{ij}, R_2 \chi_2^{ij}, R_3 \chi_2^{ij}, S_1 \chi_2^{ij}, S_2 \chi_2^{ij}\}$$

$$F_{22} = \bar{\delta}(1 - \bar{\delta})F_{33} = -diag\{\frac{R_1}{\tau_0^2}, \frac{R_2}{(\tau_1 - \tau_0)^2}, \frac{R_3}{(\tau_2 - \tau_1)^2}, \frac{S_1}{\tau_1^2}, \frac{S_2}{\tau_2^2}\}$$

with

$$\Xi_{11} = P\bar{A}_i + \bar{A}_i^T P + Q_1 + Q_2 + Q_3 - R_1 - S_1 - S_2$$

$$\Xi_{21} = R_1, \Xi_{22} = -Q_1 - R_1 - R_2, \Xi_{31} = \bar{\delta}K_j^T \bar{B}_i^T P + S_1, \Xi_{32}(\lambda) = \lambda R_2$$

$$\Xi_{33}(\lambda) = -S_1 - 2\lambda R_2, \Xi_{42}(\lambda) = (1 - \lambda)R_2, \Xi_{43}(\lambda) = \lambda R_2$$

$$\Xi_{44} = -Q_2 - R_2 - R_3, \Xi_{51} = (1 - \bar{\delta})K_j^T \bar{B}_i^T P + S_2, \Xi_{54}(\lambda) = (1 - \lambda)R_3$$

$$\Xi_{55}(\lambda) = -S_2 - 2(1 - \lambda)R_3, \Xi_{64}(\lambda) = \lambda R_3$$

$$\Xi_{65}(\lambda) = (1 - \lambda)R_3, \Xi_{66} = -Q_3 - R_3$$

$$\chi_1^{ij} = [\bar{A}_i, 0, \bar{\delta}\bar{B}_i K_j, 0, (1 - \bar{\delta})\bar{B}_i K_j, 0], \chi_2^{ij} = [0, 0, \bar{B}_i K_j, 0, -\bar{B}_i K_j, 0]$$

Proof Construct a Lyapunov–Krasovskii functional candidate as

$$V(x_t) = V_1(x_t) + V_2(x_t) + V_3(x_t) \tag{4.23}$$

where

$$V_1(x_t) = x^T(t)Px(t) \tag{4.24}$$

$$V_2(x_t) = \int_{t-\tau_0}^{t} x^T(s)Q_1x(s)ds + \int_{t-\tau_1}^{t} x^T(s)Q_2x(s)ds$$

$$+ \int_{t-\tau_2}^{t} x^T(s)Q_3x(s)ds \tag{4.25}$$

$$V_3(x_t) = \int_{-\tau_0}^{0} \int_{t+s}^{t} \dot{x}^T(v)\tau_0 R_1\dot{x}(v)dvds$$

$$+ (\tau_1 - \tau_0) \int_{-\tau_1}^{-\tau_0} \int_{t+s}^{t} \dot{x}^T(v)R_2\dot{x}(v)dvds$$

$$+ (\tau_2 - \tau_1) \int_{-\tau_2}^{-\tau_1} \int_{t+s}^{t} \dot{x}^T(v)R_3\dot{x}(v)dvds$$

$$+ \sum_{j=1}^{2} \int_{-\tau_j}^{0} \int_{t+s}^{t} \dot{x}^T(v)\tau_j S_j\dot{x}(v)dvds \tag{4.26}$$

where P, Q_i, R_i, S_1, $S_2 > 0$ $(i = 1, 2, 3)$ are to be determined. Then, the mathematical expectation of the generator $\mathscr{L}V(x_t)$ for the evolution of $V(x_t)$ along the solutions of system (4.20) is given by

$$\mathbb{E}\{\mathscr{L}V_1(x_t)\} = 2\mathbb{E}\{x^T(t)P\dot{x}(t)\} = 2x^T(t)P\varphi(t) \tag{4.27}$$

$$\mathbb{E}\{\mathscr{L}V_2(x_t)\} = x^T(t)\sum_{i=1}^{3} Q_ix(t) - \sum_{i=0}^{2} x^T(t-\tau_i)Q_{i+1}x(t-\tau_i) \tag{4.28}$$

$$\mathbb{E}\{\mathscr{L}V_3(x_t)\} = \mathbb{E}\{\dot{x}^T(t)\Lambda\dot{x}(t) - \int_{t-\tau_0}^{t} \dot{x}^T(v)\tau_0 R_1\dot{x}(v)dv$$

$$- \sum_{i=1}^{2} [\int_{t-\tau_i}^{t} \dot{x}^T(v)\tau_i S_i\dot{x}(v)dv$$

$$+ \int_{t-\tau_i}^{t-\tau_{i-1}} \dot{x}^T(v)(\tau_i - \tau_{i-1})R_{i+1}\dot{x}(v)dv]\} \tag{4.29}$$

where

$$\Lambda = \tau_0^2 R_1 + (\tau_1 - \tau_0)^2 R_2 + (\tau_2 - \tau_1)^2 R_3 + \sum_{j=1}^{2} \tau_j^2 S_j.$$

From (4.20), we have

$$\mathbb{E}\{\dot{x}^T(t)\Lambda\dot{x}(t)\} = \mathbb{E}\{[\varphi(t) + (\delta(t) - \bar{\delta})\psi(t)]^T \Lambda [\varphi(t) + (\delta(t) - \bar{\delta})\psi(t)]\}$$
$$= \varphi^T(t)\Lambda\varphi(t) + \bar{\delta}(1 - \bar{\delta})\psi^T(t)\Lambda\psi(t) \qquad (4.30)$$

Next, we prove Theorem 4.1 in two steps. First, we prove that Theorem 4.1 holds for $\tau_0 \le \tau_1(t) < \tau_1$. Second, we prove that it also holds for $\tau_1 \le \tau_2(t) < \tau_2$. One then can conclude that Theorem 4.1 is true for $\tau_0 \le \tau(t) < \tau_2$.

Case 1: $\tau_0 \le \tau_1(t) < \tau_1$.

Applying Lemmas 2.2 and 2.5 in Chap. 2 to deal with the cross product terms in Eq. (4.29), when $R_i, S_j > 0$:

$$-\int_{t-\tau_0}^{t} \dot{x}^T(v)\tau_0 R_1\dot{x}(v)dv \le \begin{bmatrix} x(t) \\ x(t-\tau_0) \end{bmatrix}^T \mathscr{X}(R_1) \begin{bmatrix} x(t) \\ x(t-\tau_0) \end{bmatrix}$$

$$-\int_{t-\tau_j}^{t} \dot{x}^T(v)\tau_j S_j\dot{x}(v)dv \le \begin{bmatrix} x(t) \\ x(t-\tau_j(t)) \end{bmatrix}^T \mathscr{X}(S_j) \begin{bmatrix} x(t) \\ x(t-\tau_j(t)) \end{bmatrix}$$

$$-\int_{t-\tau_1}^{t-\tau_0} \dot{x}^T(v)(\tau_1 - \tau_0)R_2\dot{x}(v)dv \le \begin{bmatrix} x(t-\tau_0) \\ x(t-\tau_1(t)) \\ x(t-\tau_1) \end{bmatrix}^T \mathscr{Y}(R_2) \begin{bmatrix} x(t-\tau_0) \\ x(t-\tau_1(t)) \\ x(t-\tau_1) \end{bmatrix}$$

$$-\int_{t-\tau_2}^{t-\tau_1} \dot{x}^T(v)(\tau_2 - \tau_1)R_3\dot{x}(v)dv \le \begin{bmatrix} x(t-\tau_1) \\ x(t-\tau_2) \end{bmatrix}^T \mathscr{X}(R_3) \begin{bmatrix} x(t-\tau_1) \\ x(t-\tau_2) \end{bmatrix}$$

Case 2: $\tau_1 \le \tau_2(t) < \tau_2$.

Applying the same method used in Case 1 to deal with cross product terms in Eq. (4.29) leads to:

$$-\int_{t-\tau_2}^{t-\tau_1} \dot{x}^T(v)(\tau_2 - \tau_1)R_3\dot{x}(v)dv < \begin{bmatrix} x(t-\tau_1) \\ x(t-\tau_2(t)) \\ x(t-\tau_2) \end{bmatrix}^T \mathscr{Y}(R_3) \begin{bmatrix} x(t-\tau_1) \\ x(t-\tau_2(t)) \\ x(t-\tau_2) \end{bmatrix}$$
$$(4.31)$$

$$-\int_{t-\tau_1}^{t-\tau_0} \dot{x}^T(v)(\tau_1 - \tau_0)R_2\dot{x}(v)dv \le \begin{bmatrix} x(t-\tau_0) \\ x(t-\tau_1) \end{bmatrix}^T \mathscr{X}(R_2) \begin{bmatrix} x(t-\tau_0) \\ x(t-\tau_1) \end{bmatrix} \quad (4.32)$$

Considering Case 1, Case 2, and (4.23)–(4.32) together, we have

$$\mathbb{E}\{\mathcal{L}V(x_t)\} \leq \sum_{i=1}^{r}\sum_{j=1}^{r} \rho_j \mu_i \mu_j \zeta^T(t)[F_{11}^{ij} - \Gamma_{ij}]\zeta(t)$$

$$= \sum_{i=1}^{r-1}\sum_{j>i}^{r} \rho_j \mu_i \mu_j \zeta^T(t)[F_{11}^{ij} + \frac{\rho_i}{\rho_j}F_{11}^{ji} - \Gamma_{ij} - \frac{\rho_i}{\rho_j}\Gamma_{ji}]\zeta(t)$$

$$+ \sum_{i=1}^{r} \rho_i \mu_i^2 \zeta^T(t)[F_{11}^{ii} - \Gamma_{ii}]\zeta(t) \tag{4.33}$$

where $\zeta(t) = col\{x(t), x(t-\tau_0), x(t-\tau_1(t)), x(t-\tau_1), x(t-\tau_2(t)), x(t-\tau_2)\}$, $\Gamma_{ij} = (F_{21}^{ij})^T F_{22}^{-1} F_{21}^{ij} + (F_{31}^{ij})^T F_{33}^{-1} F_{33}^{ij}$ with $F_{11}^{ij}, F_{21}^{ij}, F_{22} F_{33}^{ij}, F_{33}$ being defined in Theorem 4.1, and $\lambda = 0$ or 1.

Using Schur Complement (Lemma 2.1 in Chap. 2), (4.21) and (4.22) imply that

$$F_{11}^{ll} - \Gamma_{ll} < 0 \tag{4.34}$$

$$F_{11}^{ij} + \lambda_1 F_{11}^{ji} - \Gamma_{ij} - \lambda_2 \Gamma_{ji} < 0 \tag{4.35}$$

$$F_{11}^{ij} + \lambda_2 F_{11}^{ji} - \Gamma_{ij} - \lambda_2 \Gamma_{ji} < 0 \tag{4.36}$$

Combining (4.35), (4.36), and $(\lambda_2 - \frac{\rho_i}{\rho_j})\Gamma_{ij} < 0$, we have

$$F_{11}^{ij} + \lambda_1 F_{11}^{ji} - \Gamma_{ij} - \frac{\rho_i}{\rho_j}\Gamma_{ji} < 0 \tag{4.37}$$

$$F_{11}^{ij} + \lambda_2 F_{11}^{ji} - \Gamma_{ij} - \frac{\rho_i}{\rho_j}\Gamma_{ji} < 0 \tag{4.38}$$

Define $\varepsilon_1 = \frac{\lambda_2 - \frac{\rho_i}{\rho_j}}{\lambda_2 - \lambda_1} \geq 0$ and $\varepsilon_2 = \frac{\frac{\rho_i}{\rho_j} - \lambda_1}{\lambda_2 - \lambda_1} \geq 0$. It follows from (4.37) and (4.38) that

$$\varepsilon_1(F_{11}^{ij} + \lambda_1 F_{11}^{ji} - \Gamma_{ij} - \frac{\rho_i}{\rho_j}\Gamma_{ji}) + \varepsilon_2(F_{11}^{ij} + \lambda_2 F_{11}^{ji} - \Gamma_{ij} - \frac{\rho_i}{\rho_j}\Gamma_{ji}) < 0 \tag{4.39}$$

which yields

$$F_{11}^{ij} + \frac{\rho_i}{\rho_j}F_{11}^{ji} - \Gamma_{ij} - \frac{\rho_i}{\rho_j}\Gamma_{ji} < 0 \tag{4.40}$$

Combining (4.34) and (4.40), we have $\mathbb{E}\{\mathcal{L}V(x_t)\} < 0$. This means that $\mathbb{E}\{\mathcal{L}V(x_t)\} < \rho\|x(t)\|^2$ for a sufficiently small $\rho > 0$ and ensures the asymptotic stability of system (4.20) in the mean square. This completes the proof. $\square$

Theorem 4.1 does not take into account system parameter uncertainties. Using a similar routine method to handle norm-bounded uncertainties [44], one can obtain the following theorem. The proof of this theorem is similar to that of Theorem 4.1 and thus is omitted here.

Theorem 4.2 *For given scalars λ_1 and λ_2, $\tau_0 \leq \tau_1 \leq \tau_2$ and $\bar{\delta}$, the system (4.20) is robustly asymptotically stable in the mean square, if there exist matrices $P > 0$, $Q_i > 0$, $R_i > 0$ S_j, $(i = 1, 2, 3, j = 1, 2)$ with appropriate dimensions, scalars $\varepsilon_{1i} > 0$ and $\varepsilon_{2i} > 0$, such that the following LMIs hold for $l, i, j = 1, \ldots, r$, $1 \leq i < j \leq r$ and $\lambda = 0, 1, v = 1, 2$:*

$$\tilde{\Pi}_{ll}(\lambda) < 0 \tag{4.41}$$

$$\tilde{\Pi}_{ij}(\lambda) + \tilde{\Pi}^v_{ji}(\lambda) < 0 \tag{4.42}$$

where

$$\tilde{\Pi}_{ij}(\lambda) = \begin{bmatrix} \tilde{F}^{ij}_{11}(\lambda) & * & * & * & * \\ \tilde{F}^{ij}_{21} & \tilde{F}_{22} & * & * & * \\ \tilde{F}^{ij}_{31} & 0 & \tilde{F}_{33} & * & * \\ F^{ij}_{41} & F^{ij}_{42} & 0 & F^i_{44} & * \\ F^{ij}_{51} & 0 & F^{ij}_{42} & 0 & F^i_{55} \end{bmatrix}$$

$$\tilde{\Pi}^v_{ij}(\lambda) = \begin{bmatrix} \lambda_v \tilde{F}^{ij}_{11}(\lambda) & * & * & * & * \\ \tilde{F}^{ij}_{21} & \frac{\tilde{F}_{22}}{\lambda_2} & * & * & * \\ \tilde{F}^{ij}_{31} & 0 & \frac{\tilde{F}_{33}}{\lambda_2} & * & * \\ F^{ij}_{41} & F^{ij}_{42} & 0 & F^i_{44} & * \\ F^{ij}_{51} & 0 & F^{ij}_{53} & 0 & F^i_{55} \end{bmatrix}$$

$$F^{ij}_{41} = \begin{bmatrix} \varepsilon_{1i} H^T_i P \ 0 & 0 & 0 & 0 & 0 \\ E_{ai} & 0 \ \bar{\delta} E_{bi} K_j \ 0 \ (1 - \bar{\delta}) E_{bi} K_j \ 0 \end{bmatrix}$$

$$\tilde{F}^{ijv}_{41} = \begin{bmatrix} \varepsilon_{1i} \lambda_v H^T_i P \ 0 & 0 & 0 & 0 & 0 \\ E_{ai} & 0 \ \bar{\delta} E_{bi} K_j \ 0 \ (1 - \bar{\delta}) E_{bi} K_j \ 0 \end{bmatrix}$$

$$F^{ij}_{42} = \begin{bmatrix} \varepsilon_{2i} H^T_i R_1 \ \varepsilon_{2i} H^T_i R_2 \ \varepsilon_{2i} H^T_i R_3 \ \varepsilon_{2i} H^T_i S_1 \ \varepsilon_{2i} H^T_i S_2 \\ 0 & 0 & 0 & 0 & 0 \end{bmatrix}$$

$$F^i_{44} = diag\{-\varepsilon_{1i} I, -\varepsilon_{1i} I\}$$

$$F^{ij}_{51} = \begin{bmatrix} 0 \ 0 & 0 & 0 & 0 & 0 \\ 0 \ 0 \ E_{bi} K_j \ 0 \ -E_{bi} K_j \ 0 \end{bmatrix}$$

$$F^{ij}_{53} = \begin{bmatrix} \varepsilon_{2i} H^T_i R_1 \ \varepsilon_{2i} H^T_i R_2 \ \varepsilon_{2i} H^T_i R_3 \ \varepsilon_{2i} H^T_i S_1 \ \varepsilon_{2i} H^T_i S_2 \\ 0 & 0 & 0 & 0 & 0 \end{bmatrix}$$

$$F^i_{55} = diag\{-\varepsilon_{2i} I, -\varepsilon_{2i} I\}$$

and $\tilde{F}^{ij}_{11}(\lambda)$, $\tilde{F}^{ij}_{21}$, $\tilde{F}^{ij}_{31}$ equal to $F^{ij}_{11}(\lambda)$, F^{ij}_{21}, F^{ij}_{31} in (4.21) and (4.22) with $\bar{A}_i$ and $\bar{B}_i$ being replaced by A_i and B_i, respectively.

4.2.2 Robust Controller Design

This section studies robust controller design for networked T-S fuzzy systems. This is a further development on the results obtained in the previous section.

Theorem 4.3 *For given scalars λ_1 and λ_2, $\tau_0 \leq \tau_1 \leq \tau_2$ and $\bar{\delta}$, the system (4.20) is robustly asymptotically stable in the mean square with feedback gains $K_j = Y_j X^{-1}$, if there exist matrices $X > 0$, $\bar{Q}_i > 0$, $\bar{R}_i > 0$ $\bar{S}_j$, $(i = 1, 2, 3, j = 1, 2)$ with appropriate dimensions, scalars $\varepsilon_{1i} > 0$ and $\varepsilon_{2i} > 0$, such that the following matrix inequalities hold for $l, i, j = 1, \ldots, r$, $1 \leq i < j \leq r$ and $\lambda = 0, 1, v = 1, 2$:*

$$\Upsilon_{ll}(\lambda) < 0 \tag{4.43}$$
$$\Upsilon_{ij}(\lambda) + \Upsilon_{ji}^{v}(\lambda) < 0 \tag{4.44}$$

where

$$\Upsilon_{ij}(\lambda) = \begin{bmatrix} \Gamma_{11}^{ij}(\lambda) & * & * & * & * \\ \Gamma_{21}^{ij} & \Gamma_{22} & * & * & * \\ \Gamma_{31}^{ij} & 0 & \Gamma_{33} & * & * \\ \Gamma_{41}^{ij} & \Gamma_{42}^{ij} & 0 & \Gamma_{44}^{i} & * \\ \Gamma_{51}^{ij} & 0 & \Gamma_{42}^{ij} & 0 & \Gamma_{55}^{i} \end{bmatrix}$$

$$\Upsilon_{ij}^{v}(\lambda) = \begin{bmatrix} \lambda_v \Gamma_{11}^{ij}(\lambda) & * & * & * & * \\ \Gamma_{21}^{ij} & \Gamma_{22} & * & * & * \\ \Gamma_{31}^{ij} & 0 & \Gamma_{33} & * & * \\ \Gamma_{41}^{ijv} & \Gamma_{42}^{ij} & 0 & \Gamma_{44}^{i} & * \\ \Gamma_{51}^{ij} & 0 & \Gamma_{53}^{ij} & 0 & \Gamma_{55}^{i} \end{bmatrix}$$

and

$$\Gamma_{11}^{ij} = \begin{bmatrix} \Pi_{11} & * & * & * & * & * \\ \Pi_{21} & \Pi_{22} & * & * & * & * \\ \Pi_{31} & \Pi_{32}(\lambda) & \Pi_{33}(\lambda) & * & * & * \\ 0 & \Pi_{42}(\lambda) & \Pi_{43}(\lambda) & \Pi_{44}(\lambda) & * & * \\ \Pi_{51}(\lambda) & 0 & 0 & \Pi_{54}(\lambda) & \Pi_{55}(\lambda) & * \\ 0 & 0 & 0 & \Pi_{64}(\lambda) & \Pi_{65}(\lambda) & \Pi_{66} \end{bmatrix}$$

$$\Gamma_{21}^{ij} = col\{\bar{\chi}_1^{ij}, \bar{\chi}_1^{ij}, \bar{\chi}_1^{ij}, \bar{\chi}_1^{ij}, \bar{\chi}_1^{ij}\}, \Gamma_{31}^{ij} = col\{\bar{\chi}_2^{ij}, \bar{\chi}_2^{ij}, \bar{\chi}_2^{ij}, \bar{\chi}_2^{ij}, \bar{\chi}_2^{ij}\}$$

$$\Gamma_{22} = -diag\{\frac{X\bar{R}_1^{-1}X}{\tau_0^2}, \frac{X\bar{R}_2^{-1}X}{(\tau_1 - \tau_0)^2}, \frac{X\bar{R}_3^{-1}X}{(\tau_2 - \tau_1)^2}, \frac{X\bar{S}_1^{-1}X}{\tau_1^2}, \frac{X\bar{S}_2^{-1}X}{\tau_2^2}\}$$

$$\Gamma_{33} = (\bar{\delta}(1 - \bar{\delta}))^{-1}\Gamma_{22}$$

$$\Gamma_{41}^{ij} = \begin{bmatrix} \varepsilon_{1i}H_i^T & 0 & 0 & 0 & 0 & 0 \\ E_{ai}X & 0 & \bar{\delta}E_{bi}Y_j & 0 & (1 - \bar{\delta})E_{bi}Y_j & 0 \end{bmatrix}$$

$$\Gamma_{41}^{ijv} = \begin{bmatrix} \varepsilon_{1i}\lambda_v H_i^T & 0 & 0 & 0 & 0 & 0 \\ E_{ai}X & 0 & \bar{\delta}E_{bi}Y_j & 0 & (1-\bar{\delta})E_{bi}Y_j & 0 \end{bmatrix}$$

$$\Gamma_{42}^{ij} = \begin{bmatrix} \varepsilon_{1i}H_i^T & \varepsilon_{1i}H_i^T & \varepsilon_{1i}H_i^T & \varepsilon_{1i}H_i^T & \varepsilon_{1i}H_i^T \\ 0 & 0 & 0 & 0 & 0 \end{bmatrix}$$

$$\Gamma_{44}^i = diag\{-\varepsilon_{1i}I, -\varepsilon_{1i}I\}$$

$$\Gamma_{51}^{ij} = \begin{bmatrix} 0 & 0 & 0 & 0 & 0 & 0 \\ 0 & 0 & E_{bi}K_j & 0 & -E_{bi}Y_j & 0 \end{bmatrix}$$

$$\Gamma_{53}^{ij} = \begin{bmatrix} \varepsilon_{2i}H_i^T & \varepsilon_{2i}H_i^T & \varepsilon_{2i}H_i^T & \varepsilon_{2i}H_i^T & \varepsilon_{2i}H_i^T \\ 0 & 0 & 0 & 0 & 0 \end{bmatrix}$$

$$\Gamma_{55}^i = diag\{-\varepsilon_{2i}I, -\varepsilon_{2i}I\}$$

with

$$\Pi_{11} = A_i X + X A_i^T + \bar{Q}_1 + \bar{Q}_2 + \bar{Q}_3 - \bar{R}_1 - \bar{S}_1 - \bar{S}_2$$

$$\Pi_{21} = \bar{R}_1, \Pi_{22} = -\bar{Q}_1 - \bar{R}_1 - \bar{R}_2, \Pi_{31} = \bar{\delta}Y_j^T B_i^T + \bar{S}_1$$

$$\Pi_{32}(\lambda) = \lambda\bar{R}_2, \Pi_{33}(\lambda) = -\bar{S}_1 - 2\lambda\bar{R}_2, \Pi_{42}(\lambda) = (1-\lambda)\bar{R}_2$$

$$\Pi_{43}(\lambda) = \lambda\bar{R}_2, \Pi_{44} = -\bar{Q}_2 - \bar{R}_2 - \bar{R}_3, \Pi_{51} = (1-\bar{\delta})Y_j^T B_i^T + \bar{S}_2$$

$$\Pi_{54}(\lambda) = (1-\lambda)\bar{R}_3, \Pi_{55}(\lambda) = -\bar{S}_2 - 2(1-\lambda)\bar{R}_3$$

$$\Pi_{64}(\lambda) = \lambda\bar{R}_3, \Pi_{65}(\lambda) = (1-\lambda)\bar{R}_3, \Pi_{66} = -\bar{Q}_3 - \bar{R}_3$$

$$\bar{\chi}_1^{ij} = [A_i X, 0, \bar{\delta}B_i Y_j, 0, (1-\bar{\delta})B_i Y_j, 0], \bar{\chi}_2^{ij} = [0, 0, B_i Y_j, 0, -B_i Y_j, 0]$$

Proof Define $X = P^{-1}$, $X Q_i X = \bar{R}_i$, $X R_i X = \bar{R}_i$, $X S_j X = \bar{S}_j$ ($i = 1, 2, 3, j = 1, 2$), and $Y_j = K_j X$. Pre and postmultiply both sides of (4.41) and (4.42) with $diag\{\mathscr{J}_1, \mathscr{J}_2, \mathscr{J}_2, \mathscr{J}_3\}$ and its transpose, where $\mathscr{J}_1 = diag\{X, X, X, X, X, X\}$, $\mathscr{J}_2 = diag\{R_1^{-1}, R_2^{-1}, R_3^{-1}, S_1^{-1}, S_2^{-1}\}$, $\mathscr{J}_3 = diag\{I, I, I, I\}$, then we have (4.47) and (4.48). This completes the proof. $\square$

Theorem 4.3 cannot be directly implemented by standard numerical software due to the nonlinear terms $X\bar{R}_i^{-1}X$ and $X\bar{S}_i^{-1}X$. Since $X > 0$:

$$(\bar{R}_i - X)\bar{R}_i^{-1}(\bar{R}_i - X) > 0, \quad (\bar{S}_i - X)\bar{S}_i^{-1}(\bar{S}_i - X) > 0 \qquad (4.45)$$

which are equivalent to

$$-X\bar{R}_i^{-1}X < \bar{R}_i - 2X, \quad -X\bar{S}_i^{-1}X < \bar{S}_i - 2X \qquad (4.46)$$

By combining Theorem 4.3 and Eq. (4.46), we readily obtain the following theorem.

Theorem 4.4 *For given scalars λ_1 and λ_2, $\tau_0 \le \tau_1 \le \tau_2$ and $\bar{\delta}$, the system (4.20) is robustly asymptotically stable in the mean square with feedback gains $K_j = Y_j X^{-1}$, if there exist matrices $X > 0$, $\bar{Q}_i > 0$, $\bar{R}_i > 0$ $\bar{S}_j$, ($i = 1, 2, 3, j = 1, 2$) with* *appropriate dimensions, scalars $\varepsilon_{1i} > 0$ and $\varepsilon_{2i} > 0$, such that the following LMIs hold for $l, i, j = 1, \ldots, r$, $1 \le i < j \le r$ and $\lambda = 0, 1$, $v = 1, 2$:*

$$\tilde{\Upsilon}_{ll}(\lambda) < 0 \tag{4.47}$$

$$\tilde{\Upsilon}_{ij}(\lambda) + \tilde{\Upsilon}_{ji}^{v}(\lambda) < 0 \tag{4.48}$$

where

$$\tilde{\Upsilon}_{ij}(\lambda) = \begin{bmatrix} \Gamma_{11}^{ij}(\lambda) & * & * & * & * \\ \Gamma_{21}^{ij} & \tilde{\Gamma}_{22} & * & * & * \\ \Gamma_{31}^{ij} & 0 & \tilde{\Gamma}_{33} & * & * \\ \Gamma_{41}^{ij} & \Gamma_{42}^{ij} & 0 & \Gamma_{44}^{i} & * \\ \Gamma_{51}^{ij} & 0 & \Gamma_{42}^{ij} & 0 & \Gamma_{55}^{i} \end{bmatrix}$$

$$\tilde{\Upsilon}_{ij}^{v}(\lambda) = \begin{bmatrix} \lambda_v \Gamma_{11}^{ij}(\lambda) & * & * & * & * \\ \Gamma_{21}^{ij} & \tilde{\Gamma}_{22} & * & * & * \\ \Gamma_{31}^{ij} & 0 & \tilde{\Gamma}_{33} & * & * \\ \Gamma_{41}^{ijv} & \Gamma_{42}^{ij} & 0 & \Gamma_{44}^{i} & * \\ \Gamma_{51}^{ij} & 0 & \Gamma_{53}^{ij} & 0 & \Gamma_{55}^{i} \end{bmatrix}$$

$$\tilde{\Gamma}_{22} = diag\{\frac{\bar{R}_1 - 2X}{\tau_0^2}, \frac{\bar{R}_2 - 2X}{(\tau_1 - \tau_0)^2}, \frac{\bar{R}_3 - 2X}{(\tau_2 - \tau_1)^2}, \frac{\bar{S}_1 - 2X}{\tau_1^2}, \frac{\bar{S}_2 - 2X}{\tau_2^2}\}$$

$$\tilde{\Gamma}_{33} = (\bar{\delta}(1 - \bar{\delta}))^{-1}\tilde{\Gamma}_{22}$$

and other block matrices are defined in Theorem 4.3.

For the method employed in Theorem 4.4 to deal with the original nonconvex feasibility problem formulated in Theorem 4.3, there is an alternate algorithm based on the cone complementarity numerical approach (CCL) [96]. Although the CCL results are slightly less conservative than those based on Eq. (4.46), it needs more auxiliary variables in solving LMIs. If the number of T-S fuzzy rules is high, the extra computation load is very significant. Therefore, the algorithm based on Eq. (4.46) is used in the design of this chapter.

4.3 Numerical Examples

To demonstrate the effectiveness of the proposed approach, this section gives two practical engineering examples: (a) a mass–spring system [103]; and (b) a flexible joint robot arm system [104].

Example 1: Consider the following nonlinear mass–spring system [103]

$$\dot{x}_1 = x_2 \tag{4.49}$$
$$\dot{x}_2 = -0.01x_1 - 0.67x_1^3 + u \tag{4.50}$$

Choose fuzzy membership function as $\mu_1(x_1) = 1 - x_1^2$ and $\mu_2(x_1) = 1 - \mu_1(x_1)$, where $\mu_1 \in [-1, 1]$. Similar to [103], the following fuzzy model is also used to model aforementioned nonlinear system:

$$R^1 : \text{If } x_1(t) \text{ is } \mu_1$$
$$\text{Then } \dot{x}(t) = A_1 x(t) + B_1 u(t) \tag{4.51}$$

$$R^2 : \text{If } x_1(t) \text{ is } \mu_1$$
$$\text{Then } \dot{x}(t) = A_2 x(t) + B_2 u(t) \tag{4.52}$$

where

$$A_1 = \begin{bmatrix} 0 & 1 \\ -0.01 & 0 \end{bmatrix}, \quad A_2 = \begin{bmatrix} 0 & 1 \\ -0.68 & 0 \end{bmatrix}, \quad B_1 = B_2 = \begin{bmatrix} 0 \\ 1 \end{bmatrix}$$

Considering the system (4.20) with asynchronous premise constraints, this is out of the field of the paper in [103]. Combining the MATLAB/LMI Toolbox with the CCL algorithm [96] to solve the non-LMIs in Theorem 4.4 with $\delta = 0.8$, $\tau_0 = 0.01$, $\tau_1 = 0.4$ and different λ_1 and λ_2, as listed in Table 4.1, we have the corresponding maximum allowable upper bound τ_2 and the controller gains K_1, K_2 for asynchronous parallel distribution compensation and synchronous parallel distribution compensation methods, respectively. One can see from Table 4.1 that, the obtained maximum allowable bound τ_2 from the synchronous parallel distribution compensation method is larger than that obtained from the asynchronous parallel distribution compensation method, and is smaller than that obtained without considering the effect of a communication network (that is, $\lambda_1 = \lambda_2 = 1$ in Table 4.1). Moreover, one can see from Table 4.1 that the identical controller gains K_1 and K_2 are obtained based on the asynchronous parallel distribution compensation method. In this case, the asynchronous parallel distribution compensation controller equals to a linear controller. Furthermore, when $\lambda_1 = \lambda_2 = 1$, we have $\mu_j(\theta(t)) = \mu_j(\theta(t_k))$. This implies that no communication delay is considered in the communications, that is, system (4.20) is simplified as a point-to-point connected system.

Example 2: The flexible joint robot arm model is given as [104]

$$(I_1 + \delta I_1)\ddot{\theta}_1 + (mgl + \delta m)\sin(\theta_1) + (k + \delta k)(\theta_1 - \theta_2) = 0$$
$$(I_2 + \delta I_2)\ddot{\theta}_2 + (b + \delta b)\dot{\theta}_2 + (k + \delta k)(\theta_2 - \theta_1) = u + \delta u \tag{4.53}$$

where

$$|\delta I_1| \le cI_1, |\delta I_2| \le cI_2, |\delta m| \le cmgl$$
$$|\delta k| \le ck, |\delta u| \le cu, |\delta b| \le cb.$$

Table 4.1 Upper delay bound τ_2 and controller feedback gains for different values of λ_1, λ_2 and given $\bar{\delta} = 0.8$, $\tau_0 = 0.01$, $\tau_1 = 0.4$

0. λ_1, λ_2	τ_2	$(K_1^T, K_2^T)^T$	Comments
–, –	0.58	$\begin{pmatrix} -0.3596 & -1.1915 \\ -0.3596 & -1.1915 \end{pmatrix}$	Linear K_1, K_2
0.6, 1.4	0.70	$\begin{pmatrix} -0.3631 & -1.1876 \\ -0.1183 & -1.1394 \end{pmatrix}$	Different K_1, K_2
0.8, 1.2	0.79	$\begin{pmatrix} -0.3450 & -1.1736 \\ -0.0087 & -1.0684 \end{pmatrix}$	Different K_1, K_2
1, 1	0.84	$\begin{pmatrix} -0.3210 & -1.2429 \\ 0.0609 & -1.0162 \end{pmatrix}$	Without network

Set $x_1 = \theta_1$, $x_2 = \dot{\theta}_1$, $x_3 = \theta_2$, $x_4 = \dot{\theta}_2$, then the state equation of the system is

$$\dot{x}(t) = \begin{bmatrix} 0 & 1 & 0 & 0 \\ \rho_1 & 0 & \rho_2 & 0 \\ 0 & 0 & 0 & 1 \\ \rho_3 & 0 & -\rho_3 & \rho_4 \end{bmatrix} x(t) + \begin{bmatrix} 0 \\ 0 \\ 0 \\ \frac{1+\delta}{I_2+\delta I_2} \end{bmatrix} u(t) \qquad (4.54)$$

where

$$\rho_1 = -\frac{mgl + \delta m}{I_1 + \delta I_1} \frac{\sin x_1(t)}{x_1(t)} - \rho_2, \rho_2 = \frac{k + \delta k}{I_1 + \delta I_1}$$

$$\rho_3 = \frac{k + \delta k}{I_2 + \delta I_2}, \rho_4 = -\frac{b + \delta b}{I_2 + \delta I_2}$$

$$x(t) = \begin{bmatrix} x_1(t) & x_2(t) & x_3(t) & x_4(t) \end{bmatrix}^T$$

Similar to [44], the system parameters are: $m = 0.01$ kg, $I_1 = I_2 = 1$ kgm^2, $k = 0.05$ N.m/rad, $l = 1$ m, $b = 0.007$ N.ms/rad, $c = 10\%$, $g = 9.81$ m/s^2. In line with [104], the following uncertain time-delay T-S model is established for system (4.54):

$$R^1 : \text{If } x_1(t) \text{ is } 0$$
$$\text{Then } \dot{x}(t) = [A_1 + H_1 F_1 E_{a1}]x(t) + [B_1 + H_1 F_1 E_{b1}]u(t) \qquad (4.55)$$

$$R^2 : \text{If } x_1(t) \text{ is } \pm \pi/2$$
$$\text{Then } \dot{x}(t) = [A_2 + H_2 F_2 E_{a2}]x(t) + [B_2 + H_2 F_2 E_{b2}]u(t) \qquad (4.56)$$

Table 4.2 Upper delay bounds τ_2 and controller feedback gains for $\tau_0 = 0.01$ s, $\delta(t) = 1$, and $\lambda_1 = \lambda_2 = 1$

Method	τ_2	$(K_1^T, K_2^T)^T$
Jiang and Han [44]	0.8441	$\begin{pmatrix} 0.5272 \ 0.0604 \ -0.2735 \ -0.8792 \\ 0.5333 \ 0.0816 \ -0.2696 \ -0.8768 \end{pmatrix}$
Theorem 4.4	0.8497	$\begin{pmatrix} 0.2426 \ 0.4843 \ -0.1160 \ -0.6678 \\ 0.2495 \ 0.4759 \ -0.1172 \ -0.6666 \end{pmatrix}$

where

$$A_1 = \begin{bmatrix} 0 & 1 & 0 & 0 \\ -0.1481 & 0 & 0.05 & 0 \\ 0 & 0 & 0 & 1 \\ 0.05 & 0 & -0.05 & -0.007 \end{bmatrix}$$

$$A_2 = \begin{bmatrix} 0 & 1 & 0 & 0 \\ -0.1125 & 0 & 0.05 & 0 \\ 0 & 0 & 0 & 1 \\ 0.05 & 0 & -0.05 & -0.007 \end{bmatrix}$$

$$B_i = \begin{bmatrix} 0 \\ 0 \\ 0 \\ 1 \end{bmatrix}, H_i = \begin{bmatrix} 1 & 0 & 0 & 0 \\ 0 & 1 & 0 & 0 \\ 0 & 0 & 1 & 0 \\ 0 & 0 & 0 & 1 \end{bmatrix}$$

$$E_{ai} = \begin{bmatrix} 0 & 0 & 0 & 0 \\ 2/9 & 0 & 2/9 & 0 \\ 0 & 0 & 0 & 1 \\ 2/9 & 0 & 2/9 & 2/9 \end{bmatrix}, E_{bi} = \begin{bmatrix} 0 \\ 0 \\ 0 \\ 2/9 \end{bmatrix}.$$

Table 4.2 gives the results of the two designs; one is based on the method given in Jiang and Han [44] and the other is based on Theorem 4.4 of this chapter. The same lower delay bound $\tau_0 = 0.01$ s is used in both the designs. As a simple case, here all delay is assumed within the range of $[\tau_0, \tau_2)$ $(\delta(t) \equiv 1)$ without considering its distribution. The delay upper bound τ_2 is the result of a particular design. A big value of τ_2 means a less conservative design. From Table 4.2 one can see that the method given in this paper is less conservative than that of [44].

Now considering that the network-induced delay is nonuniformly distributed and has a multifractal nature, this is beyond the scope of the design method in [44]. The results of the designs based on the method in this chapter are given in Table 4.3, where $\bar{\delta} = 0.8$ and $\tau_0 = 0.01$ s. Compare the results in Tables 4.2 and 4.3, the values of the upper bound τ_2 in Table 4.3 are higher than those in Table 4.2. This demonstrates that the design method developed in this chapter is indeed better in terms of less conservative results.

Table 4.3 Upper delay
bounds τ_2 and controller
feedback gains for different
values of τ_1 and given
$\bar{\delta} = 0.8, \lambda_1 = \lambda_2 = 1$

τ_1	τ_2	$(K_1^T, K_2^T)^T$
0.4	1.33	$\begin{pmatrix} 0.3202 \; 0.6112 \; -0.1425 \; -0.7572 \\ 0.3320 \; 0.5751 \; -0.1503 \; -0.7617 \end{pmatrix}$
0.5	1.31	$\begin{pmatrix} 0.2990 \; 0.6119 \; -0.1332 \; -0.7393 \\ 0.3102 \; 0.5786 \; -0.1395 \; -0.7385 \end{pmatrix}$
0.6	1.25	$\begin{pmatrix} 0.3012 \; 0.5719 \; -0.1366 \; -0.7379 \\ 0.3115 \; 0.5413 \; -0.1427 \; -0.7386 \end{pmatrix}$
0.7	1.18	$\begin{pmatrix} 0.2872 \; 0.5694 \; -0.1307 \; -0.7242 \\ 0.2955 \; 0.5472 \; -0.1347 \; -0.7277 \end{pmatrix}$
0.8	1.07	$\begin{pmatrix} 0.2787 \; 0.5486 \; -0.1283 \; -0.7139 \\ 0.2867 \; 0.5306 \; -0.1317 \; -0.7134 \end{pmatrix}$

4.4 Conclusion

Delay distribution-dependent robust stabilization for a class of networked T-S fuzzy
system with asynchronous premise constraints is studied in the chapter, where the
random delay is caused by the networked communication and has an interval distribution, and the premises of plant and the fuzzy rules are asynchronous. Simplified
and improved delay distribution-dependent stability criteria are achieved by using
an improved NCS model. Since more detailed delay characteristics are employed
and synchronous premises between the plant and the controller are constructed in
the development of the stability criteria, the results obtained are less conservative
than those achieved in some existing results. Two numerical examples are used to
demonstrate the effectiveness of the theoretical results presented.

Chapter 5
Decentralized Control for IP-based Large-Scale Systems

This chapter addresses the decentralized control for a large-scale system with an IP-based communication network. The decentralized controller design specifically takes the nonuniform IP-based network delay distribution characteristic into account. First, a networked decentralized control modeling for the large-scale system is proposed. The designed controller does not depend on the full-order state of the system, and considers the nonuniform network delay distribution characteristic of IP-based communication network. Second, under the assumption of the probability distribution of IP-based network, delay is known a priori, sufficient stability and stabilization conditions for the networked large-scale system are derived in terms of linear matrix inequalities. The solvability of the design depends both on the probability distribution of the communication delay and on the network topology. Finally, the design method is applied to two pendulums coupled by a spring and a quadruple-tank process [105].

This chapter is organized as follows. Section 5.1 presents a decentralized control model for a large-scale system to incorporate the specific characteristics of IP-based communication delays. Section 5.2 investigates the stability and stabilization for the system modeled in Sect. 5.1. Two numerical examples are given in Sect. 5.3. Finally, Sect. 5.4 concludes the chapter.

5.1 System and Problem Descriptions

5.1.1 Modeling of Large-Scale Systems

Consider a large-scale system [105–107]

$$S : \dot{x}(t) = Ax(t) + Bu(t) + g(t, x(t)) + f(t, x(t), u(t)) \tag{5.1}$$

© Springer-Verlag Berlin Heidelberg 2015
C. Peng et al., *Communication and Control for Networked Complex Systems*,
DOI 10.1007/978-3-662-46813-5_5

which is an interconnection of N subsystems

$$S_i : \dot{x}_i(t) = A_i x_i(t) + B_i u_i(t) + g_i(t, x_i(t)), \quad i = 1, 2, \ldots, N \qquad (5.2)$$

where $x_i(t) \in \mathbb{R}^{n_i}$ is the state and $u_i(t) \in \mathbb{R}^{m_i}$ is the input of (5.2), and A_i and B_i are constant matrices of appropriate dimensions, which constitute stabilizable pairs (A_i, B_i). In (5.1), $x(t) = (x_1^T(t), x_2^T(t), \ldots, x_N^T(t))^T \in \mathbb{R}^n$ is the state; $u(t) = (u_1^T(t), u_2^T(t), \ldots, u_N^T(t))^T \in \mathbb{R}^m$ is the input of the interconnected system (5.1), where $n = \sum_{i=1}^{N} n_i, m = \sum_{i=1}^{N} m_i$; $g : \mathbb{R} \times \mathbb{R}^n \to \mathbb{R}^n$ and $f : \mathbb{R} \times \mathbb{R}^n \times \mathbb{R}^m \to \mathbb{R}^n$ express nonlinear functions. The system matrices in (5.1) are defined as $A = diag\{A_1, A_2, \ldots, A_N\}$, $B = diag\{B_1, B_2, \ldots, B_N\}$. In (5.2), the $g_i(t, x_i(t))$ represents the structured uncertainty and is independent of other subsystem's information. Similar to [106, 107], we assume that $g_i(t, x_i(t))$ satisfies

$$\|g_i(t, x_i(t))\|^2 \le \|G_i x_i(t)\|^2, \quad \text{for all}(t, x_i(t)) \in \mathbb{R} \times \mathbb{R}^{n_i} \qquad (5.3)$$

where $g_i : \mathbb{R} \times \mathbb{R}^{n_i} \to \mathbb{R}^{n_i}$ is a vector component of $g = (g_1^T, g_2^T, \ldots, g_N^T)^T$ and $G_i \in \mathbb{R}^{n_i \times n_i}$.

For the decoupled nonlinear function $g(t, x(t))$, from (5.3), one can get

$$\|g(t, x(t))\|^2 \le \|Gx(t)\|^2, \quad \text{for all } (t, x(t)) \in \mathbb{R} \times \mathbb{R}^n \qquad (5.4)$$

where $G = diag\{G_1, G_2, \ldots, G_N\}$.

For the coupled interconnected function $f(t, x(t), u(t))$, it is assumed that

$$\|f(t, x(t), u(t))\|^2 \le \|H_1 x(t)\|^2 + \|H_2 u(t)\|^2 \qquad (5.5)$$

where $H_1 \in \mathbb{R}^{n \times n}$, $H_2 \in \mathbb{R}^{m \times m}$, for all $(t, x(t), u(t)) \in \mathbb{R} \times \mathbb{R}^n \times \mathbb{R}^m$.

Assume that $g(t, x(t))$ and $f(t, x(t), u(t))$ are sufficiently smooth so that the solution of (5.1) exists and is unique for all initial conditions and all fixed inputs u; $g(t, 0) = 0$, $f(t, 0, 0) = 0$, and $x = 0$ is assumed to be the unique equilibrium of S when $u = 0$. For convenience, in the sequel, g and f are used to denote $g(t, x(t))$ and $f(t, x(t), u(t))$, respectively.

Throughout this chapter, it is assumed that the system (5.1) is controlled through an IP-based network and the system state is available for feedback control, and that the Quality of Service (QoS) provided by the network is not perfect. Therefore, there are communication delays, packet dropouts, and disorder packets in the communications.

5.1.2 Delay Distribution-Dependent Modeling for an IP-based Large-Scale System

Figure 5.1 shows a networked control structure for a large-scale system (5.1), where solid lines represent physical links and broken lines for signal flows, and the

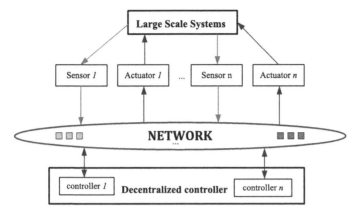

Fig. 5.1 Typical NCS setup for decentralized control

decentralized controller is adopted for the large-scale system where there are multi-sensor and multi-actuator nodes. In Fig. 5.1, each labeled "Sensor", such as "Sensor 1", represents a sensing source of a group of sensors which can send several sampled signals from one location in one data packet. In contrast to a typical NCS structure with the centralized controller in [107], the decentralized controller is employed in Fig. 5.1. Moreover, if there is only one sensing source and only one actuator node, this becomes one sensor and one actuator structure with single packet transmission in [89, 108, 109].

Under the proposed framework, the following assumptions are needed in this chapter:

- Multipacket transmission is used for the system (5.1) controlled over an IP-based network. That is, for different sensor nodes located in different places, the measurements taken there are sent via different packets.
- Each transmitted packet is time-stamped and with the identifier, where the role of identifier is used to identify the sequence number of the subsystem. The sensors are time-triggered, i.e., the system states are sampled periodically at a constant period h. Moreover, the sampling instances $t_k h$ are synchronized among the sensors in different subsystems, where t_k ($k = 1, 2, 3, \ldots$) are some integers such that $\{t_1, t_2, t_3, \ldots\} \subset \{0, 1, 2, 3, \ldots\}$.
- The decentralized controller is event-triggered. It calculates the control signal as soon as it receives a packet from its uplink. Upon the completion of the calculation, the control packet is transmitted to its downlink. Moreover, there is a buffer among the subsystems to choose and store the largest communication delay η_{t_k} in different subsystems, i.e., $\eta_{t_k} = \max v_{t_k}^i$ ($i = 1, 2, \ldots, N$), where $v_{t_k}^i$ is the communication delay in subsystem i at sampling instant $t_k h$. Communication delays in every subsystems are chosen as η_{t_k} in the modeling.

- The actuators are event-triggered with a serial of logic ZOH. The function of a logic ZOH is to select and to store the latest control signal based on the time stamps of received packets.

Based on the above assumptions, the following decentralized networked state feedback controller is designed for subsystem $S_i, i = 1, 2, \ldots, N$

$$u_i(t^+) = k_i x_i(t_k h), \quad t \in \{t_k h + \eta_{t_k}, k = 1, 2, \ldots\} \tag{5.6}$$

where $u_i(t^+) = \lim_{\hat{t} \to t+0} u_i(\hat{t})$, k_i is controller gain to be determined.
Define

$$\tau(t) = t - t_k h, \quad t \in \Omega = [t_k h + \eta_{t_k}, t_{k+1} h + \eta_{t_{k+1}}). \tag{5.7}$$

From the above assumptions, it is known that each data packet is time-stamped. Then the controller can choose the latest time-stamped packet as the input. In this sense, the condition of $t_{k+1} h + \eta_{t_{k+1}} > t_k h + \eta_{t_k}$ is guaranteed in (5.7).

Although the subsystems use the different channels of the same network to transmit the sampled signal, it is reasonable to assume that there are the identical statistic characteristics of the packets transferred in these communication channels. Also, it is known that to a specified communication network, communication delay and the number of consecutive packet losses are bounded [110]. From (5.7), we have

$$\tau_1 \leq \eta_{t_k} \leq \tau(t) < (t_{k+1} - t_k)h + \eta_{t_{k+1}} \leq \tau_3 \tag{5.8}$$

where $\tau_1 = \inf_k \{\eta_{t_k}\}$, $\tau_3 = \sup_k [(t_{k+1} - t_k)h + \eta_{t_{k+1}}]$. Since τ_3 is dependent on the communication delay and packet dropout, this enables that all kinds of nonideal network quality of service can be incorporated in an integrated maximum allowable equivalent delay bound (MAEDB) τ_3 [15].

From (5.6) and (5.7), the control output in the logic ZOH can be represented by

$$u_i(t^+) = k_i x_i(t - \tau(t)), \quad t \in \Omega, \ i = 1, 2, \ldots, N \tag{5.9}$$

From (5.9), one can see that the decentralized controller for the system (5.1) can be written as

$$u(t^+) = K x(t - \tau(t)), \quad t \in \Omega \tag{5.10}$$

where K is of the compatible block-diagonal structure; $x(t - \tau(t))$ means $col\{x_1(t - \tau(t)), \ldots, x_N(t - \tau(t))\}$.

Based on (5.1) and (5.10), the dynamics of the system is

$$\dot{x}(t) = Ax(t) + BK x(t - \tau(t)) + g + f, \quad t \in \Omega \tag{5.11}$$

To fully use this statistic characteristic of the IP-based networks and assume there is a network between the sensors and controllers, based on the delay distribution-dependent modeling method mentioned in Sect. 2.1.2, the following delay distribution-dependent feedback control law (5.12) is adopted to replace the general form of the control law given in (5.10)

$$u(t) = \delta(t)K_1 x(t - \tau_1(t)) + (1 - \delta(t))K_2 x(t - \tau_2(t)), \quad t \in \Omega \qquad (5.12)$$

where

$$\begin{aligned} \tau_1(t) &= \delta(t)\tau(t), \text{ and } \delta(t) = 1 \text{ if } \tau(t) \in [\tau_1, \tau_2) \\ \tau_2(t) &= (1 - \delta(t))\tau(t), \text{ and } \delta(t) = 0 \text{ if } \tau(t) \in [\tau_2, \tau_3) \end{aligned} \qquad (5.13)$$

The $\tau_1(t)$ and $\tau_2(t)$ in (5.12) are important in the analysis and synthesis of the networked closed-loop system (5.11). $\tau_1(t)$ means that the distribution of $\tau(t)$ is within a lower end range of $[\tau_1, \tau_2)$; while $\tau_2(t)$ means that the distribution of $\tau(t)$ is within a higher end range of $[\tau_2, \tau_3)$. Similar to [43, 85], it is also assumed that $\delta(t)$ in (5.12) is a bernoulli distributed sequence (BDS) with

$$\begin{cases} Prob\{\delta(t) = 1\} = \mathbb{E}\{\delta(t)\} := \bar{\delta} \\ Prob\{\delta(t) = 0\} = 1 - \mathbb{E}\{\delta(t)\} := 1 - \bar{\delta} \end{cases} \qquad (5.14)$$

where $0 \leq \bar{\delta} \leq 1$ is a constant and is determined by the chosen network and the NCSs configuration.

From (5.11) and (5.12), the following closed-loop system model can be obtained.

$$\begin{aligned} \dot{x}(t) = {}& Ax(t) + \delta(t)BK_1 x(t - \tau_1(t)) \\ & + (1 - \delta(t))BK_2 x(t - \tau_2(t)) + g + f, \quad t \in \Omega \end{aligned} \qquad (5.15)$$

Notice that in Fig. 5.1, the signal transfers from the sensors to the controllers and from the controllers to the actuators are not perfect. For example, there are communication delay, possible data packet dropout and/or out-of-order. However, because of the time stamp, the Logic ZOH guarantees that $t_{k+1} > t_k$. In other words, the out-of-order packets are discarded in the actuators. Moreover, notice that in (5.8) the communication delay and the delay due to the packet dropouts are included in the overall $\tau(t)$. Therefore, the data dropout and communication delay have been included in (5.10).

5.2 Stability Analysis and Controller Synthesis

In this section, an approach for stability analysis and controller synthesis of the system (5.1) is developed. First, the following stability analysis result is stated.

Theorem 5.1 *For some given constants* $0 < \tau_1 < \tau_2 < \tau_3, 0 < \bar{\delta} < 1,$ *and matrices* K_1 *and* K_2, *the system* (5.15) *is asymptotically stable in the mean square if there exist real matrices* $P > 0$, $Q_i > 0$, $R_i > 0$ $(i = 1, 2, 3)$, *scalars* $\varepsilon_i > 0$ $(i = 1, 2)$, *and matrices* U_j $(j = 2, 3)$ *with appropriate dimensions such that*

$$
\begin{bmatrix}
\Gamma_{11} & * & * & * \\
\Gamma_{21} & \Gamma_{22} & * & * \\
\Gamma_{31} & 0 & \Gamma_{33} & * \\
\Gamma_{41} & 0 & 0 & \Gamma_{44}
\end{bmatrix} < 0 \tag{5.16}
$$

$$
\Omega_j = \begin{bmatrix} R_j & * \\ U_j & R_j \end{bmatrix} \geq 0, \quad j = 2, 3 \tag{5.17}
$$

where

$$
\Gamma_{11} = \begin{bmatrix}
F_{11} & * & * & * & * & * & * & * \\
F_{21} & F_{22} & * & * & * & * & * & * \\
F_{31} & F_{32} & F_{33} & * & * & * & * & * \\
0 & F_{42} & F_{32} & F_{44} & * & * & * & * \\
F_{51} & 0 & 0 & F_{54} & F_{55} & * & * & * \\
0 & 0 & 0 & F_{64} & F_{54} & F_{66} & * & * \\
P & 0 & 0 & 0 & 0 & 0 & -\varepsilon_1 I & * \\
P & 0 & 0 & 0 & 0 & 0 & 0 & -\varepsilon_2 I
\end{bmatrix}
$$

$$
\Gamma_{21} = \begin{bmatrix} \mathscr{I}_1^T & \mathscr{I}_1^T & \mathscr{I}_1^T \end{bmatrix}^T, \quad \Gamma_{31} = \begin{bmatrix} \mathscr{I}_2^T & \mathscr{I}_2^T & \mathscr{I}_2^T \end{bmatrix}^T
$$

$$
\mathscr{I}_1 = \begin{bmatrix} A & 0 & \bar{\delta}BK_1 & 0 & (1 - \bar{\delta})BK_2 & 0 & I & I \end{bmatrix}
$$

$$
\mathscr{I}_2 = \begin{bmatrix} A & 0 & BK_1 & 0 & -BK_2 & 0 & 0 & 0 \end{bmatrix}
$$

$$
\Gamma_{22} = -diag\{\frac{R_1^{-1}}{\tau_1}, \frac{R_2^{-1}}{\tau_2}, \frac{R_3^{-1}}{\tau_3}\}, \quad \Gamma_{33} = \frac{\Gamma_{22}}{\bar{\delta}(1 - \bar{\delta})}
$$

$$
\Gamma_{41} = \begin{bmatrix}
G & 0 & 0 & 0 & 0 & 0 & 0 & 0 \\
H_1 & 0 & 0 & 0 & 0 & 0 & 0 & 0 \\
0 & 0 & \sqrt{\bar{\delta}}H_2K_1 & 0 & \Lambda & 0 & 0 & 0
\end{bmatrix}
$$

$$
\Gamma_{44} = diag\{-\varepsilon_1^{-1}I, -\varepsilon_2^{-1}I, -\varepsilon_2^{-1}I\}
$$

with

$$
F_{11} = PA + A^T P + \sum_{i=1}^{3} \frac{\tau_1 Q_i - R_i}{\tau_1}, \quad \Lambda = \sqrt{1 - \bar{\delta}}H_2K_2
$$

$$
F_{21} = \sum_{i=1}^{3} \frac{R_i}{\tau_1}, \quad F_{22} = -\sum_{i=1}^{3} \frac{R_i}{\tau_1} - \frac{R_2 + R_3}{\tau_2 - \tau_1} - Q_1
$$

$$
F_{31} = \bar{\delta}K_1^T B^T P, \quad F_{32} = \sum_{i=2}^{3} \frac{R_i - U_i}{\tau_2 - \tau_1}
$$

$$F_{33} = \sum_{i=2}^{3} \frac{U_i^T + U_i - 2R_i}{\tau_2 - \tau_1}, F_{42} = \frac{U_2 + U_3}{\tau_2 - \tau_1}$$

$$F_{44} = -Q_2 - \frac{R_2 + R_3}{\tau_2 - \tau_1} - \frac{R_3}{\tau_3 - \tau_2}, F_{51} = \frac{K_2^T B^T P}{(1 - \bar{\delta})^{-1}}$$

$$F_{54} = \frac{R_3 - U_3}{\tau_3 - \tau_2}, F_{55} = \frac{U_3^T + U_3 - 2R_3}{\tau_3 - \tau_2}$$

$$F_{64} = \frac{U_3}{\tau_3 - \tau_2}, F_{66} = -Q_3 - \frac{R_3}{\tau_3 - \tau_2}$$

Proof Construct a Lyapunov-Krasovskii functional candidate as

$$V(x_t) = x^T(t)Px(t) + \sum_{i=1}^{3} \int_{t-\tau_i}^{t} x^T(s)Q_i x(s)ds$$

$$+ \sum_{i=1}^{3} \int_{-\tau_i}^{0} \int_{t+s}^{t} \dot{x}^T(v)R_i \dot{x}(v)dvds \qquad (5.18)$$

where $P > 0$, $Q_i > 0$, $R_i > 0$ ($i = 1, 2, 3$) are to be determined, and $\tau_0 = 0$. It follows from Eq. (5.14) that $\mathbb{E}\{\delta(t) - \bar{\delta}\} = 0$, $\mathbb{E}\{(\delta(t) - \bar{\delta})^2\} = \bar{\delta}(1 - \bar{\delta})$. Then, the mathematical expectation of the generator $\mathscr{L}V(x_t)$ for the evolution of $V(x_t)$ along the solutions of system (5.15) is given by

$$\mathbb{E}\{\mathscr{L}V(x_t)\} = 2x^T(t)P\varphi(t)$$

$$+ \sum_{i=1}^{3} [x^T(t)Q_i x(t) - x^T(t - \tau_i)Q_i x(t - \tau_i)]$$

$$+ \sum_{i=1}^{3} \mathbb{E}\{\dot{x}^T(t)\tau_i R_i \dot{x}(t) - \int_{t-\tau_i}^{t} \dot{x}^T(v)R_i \dot{x}(v)dv\} \qquad (5.19)$$

From (5.15), we have

$$\sum_{i=1}^{3} \mathbb{E}\{\dot{x}^T(t)\tau_i R_i \dot{x}(t)\}$$

$$= \sum_{i=1}^{3} \mathbb{E}\{[\varphi + (\delta(t) - \bar{\delta})\psi]^T \tau_i R_i [\varphi + (\delta(t) - \bar{\delta})\psi]\}$$

$$= \sum_{i=1}^{3} [\varphi^T \tau_i R_i \varphi + \bar{\delta}(1 - \bar{\delta})\psi^T \tau_i R_i \psi] \qquad (5.20)$$

where

$$\varphi \triangleq A_D x(t) + \bar{\delta} B_D K_1 x(t - \tau_1(t)) + (1 - \bar{\delta}) B_D K_2 x(t - \tau_2(t)) + g + f$$
$$\psi \triangleq B_D K_1 x(t - \tau_1(t)) - B_D K_2 x(t - \tau_2(t))$$

The integral items in (5.19) can be written as

$$\sum_{i=1}^{3} \int_{t-\tau_i}^{t} \dot{x}^T(v) R_i \dot{x}(v) dv = \int_{t-\tau_3}^{t-\tau_2} \dot{x}^T(v) R_3 \dot{x}(v) dv$$

$$+ \int_{t-\tau_2}^{t-\tau_1} \dot{x}^T(v)(R_2 + R_3)\dot{x}(v) dv + \sum_{i=1}^{3} \int_{t-\tau_1}^{t} \dot{x}^T(v) R_i \dot{x}(v) dv \quad (5.21)$$

Applying Lemmas 2.2 and 2.7 in Chap. 2 to deal with the integral terms in (5.21), for $R_i > 0, i = 1, 2, 3$, one can get

$$-\sum_{i=1}^{3} \int_{t-\tau_1}^{t} \dot{x}^T(v) R_i \dot{x}(v) dv$$

$$\leq -\sum_{i=1}^{3} [x(t) - x(t - \tau_1)]^T \frac{R_i}{\tau_1} [x(t) - x(t - \tau_1)] \quad (5.22)$$

$$-\int_{t-\tau_2}^{t-\tau_1} \dot{x}^T(v)(R_2 + R_3)\dot{x}(v) dv \leq -\xi^T \Pi_1^T \frac{\Omega_2 + \Omega_3}{\tau_2 - \tau_1} \Pi_1 \xi \quad (5.23)$$

$$-\int_{t-\tau_3}^{t-\tau_2} \dot{x}^T(v) R_3 \dot{x}(v) dv \leq -\xi^T \Pi_2^T \frac{\Omega_3}{\tau_3 - \tau_2} \Pi_2 \xi \quad (5.24)$$

where ξ denotes $\xi(t)$ for simplicity, and

$$\xi(t) = col[x(t), \ x(t - \tau_1), \ x(t - \tau_1(t)), x(t - \tau_2), x(t - \tau_2(t)), \ x(t - \tau_3), \ g, \ f]$$

$$\Pi_1 = \begin{bmatrix} 0 & 1 & -1 & 0 & 0 & 0 & 0 & 0 \\ 0 & 0 & 1 & -1 & 0 & 0 & 0 & 0 \end{bmatrix}, \Pi_2 = \begin{bmatrix} 0 & 0 & 0 & 1 & -1 & 0 & 0 & 0 \\ 0 & 0 & 0 & 0 & 1 & -1 & 0 & 0 \end{bmatrix}$$

Considering (5.18)–(5.24) together, for any $\varepsilon_i > 0$ $(i = 1, 2)$, we have

$$\mathbb{E}\{\mathscr{L}V(x_t)) < \mathbb{E}\{\xi^T[\Gamma_{11} - \Gamma_{21}^T \Gamma_{22}^{-1} \Gamma_{21} - \Gamma_{31}^T \Gamma_{33}^{-1} \Gamma_{31}]\xi + \Delta\} \quad (5.25)$$

where Γ_{ij} $(i, j = 1, 2)$ are defined in Theorem 5.1 and

$$\xi(t) = col[x(t), \ x(t - \tau_1), \ x(t - \tau_1(t)), x(t - \tau_2), x(t - \tau_2(t)), \ x(t - \tau_3), \ g, \ f],$$
$$\Delta \triangleq \varepsilon_1 \|g(t, x(t))\|^2 + \varepsilon_2 \|f(t, x(t), u(t))\|^2$$

For (5.4) and (5.5), we have

$$\Delta \leq \varepsilon_1 \|Gx(t)\|^2 + \varepsilon_2 \|H_1 x(t)\|^2 + \varepsilon_2 \|H_2 u(t)\|^2 \tag{5.26}$$

Using Schur complements, (5.16) and (5.26) guarantee that $\mathbb{E}\{\mathscr{L}V(x_t)\} < 0$ in (5.25). This means $\mathbb{E}\{\mathscr{L}V(x_t)\} < -\rho \|x(t)\|^2$ for a sufficiently small $\rho > 0$, and ensures the asymptotic stability of system (5.15) in the mean square. This completes the proof. □

Based on Theorem 5.1, the state feedback controller can be designed for the closed-loop system (5.15).

Theorem 5.2 *For some given constants $0 < \tau_1 < \tau_2 < \tau_3$, $0 < \bar{\delta} < 1$, the system (5.15) is asymptotically stable in the mean square with feedback gains $K_i = Y_i X^{-T}$ ($i = 1, 2$) if there exist real matrices $X > 0$, $\tilde{Q}_i > 0$, $\tilde{R}_i > 0$ ($i = 1, 2, 3$), scalars $\tilde{\varepsilon}_i > 0$ ($i = 1, 2$), and matrices $\tilde{U}_j$ ($j = 2, 3$) with appropriate dimensions such that*

$$\begin{bmatrix} \tilde{\Gamma}_{11} & * & * & * \\ \tilde{\Gamma}_{21} & \tilde{\Gamma}_{22} & * & * \\ \tilde{\Gamma}_{31} & 0 & \tilde{\Gamma}_{33} & * \\ \tilde{\Gamma}_{41} & 0 & 0 & \tilde{\Gamma}_{44} \end{bmatrix} < 0 \tag{5.27}$$

$$\tilde{\Omega}_j = \begin{bmatrix} \tilde{R}_j & * \\ \tilde{U}_j & \tilde{R}_j \end{bmatrix} \geq 0, \quad j = 2, 3 \tag{5.28}$$

where

$$\tilde{\Gamma}_{11} = \begin{bmatrix} \tilde{F}_{11} & * & * & * & * & * & * & * \\ \tilde{F}_{21} & \tilde{F}_{22} & * & * & * & * & * & * \\ \tilde{F}_{31} & \tilde{F}_{32} & \tilde{F}_{33} & * & * & * & * & * \\ 0 & \tilde{F}_{42} & \tilde{F}_{32} & \tilde{F}_{44} & * & * & * & * \\ \tilde{F}_{51} & 0 & 0 & \tilde{F}_{54} & \tilde{F}_{55} & * & * & * \\ 0 & 0 & 0 & \tilde{F}_{64} & \tilde{F}_{54} & \tilde{F}_{66} & * & * \\ \sqrt{\tilde{\varepsilon}_1}I & 0 & 0 & 0 & 0 & 0 & -I & * \\ \sqrt{\tilde{\varepsilon}_2}I & 0 & 0 & 0 & 0 & 0 & 0 & -I \end{bmatrix}$$

$$\tilde{\Gamma}_{21} = \begin{bmatrix} \tilde{\mathscr{I}}_1^T & \tilde{\mathscr{I}}_1^T & \tilde{\mathscr{I}}_1^T \end{bmatrix}^T, \quad \tilde{\Gamma}_{31} = \begin{bmatrix} \tilde{\mathscr{I}}_2^T & \tilde{\mathscr{I}}_2^T & \tilde{\mathscr{I}}_2^T \end{bmatrix}^T$$

$$\tilde{\Gamma}_{22} = -diag\{\frac{XR_1^{-1}X}{\tau_1}, \frac{XR_2^{-1}X}{\tau_2}, \frac{XR_3^{-1}X}{\tau_3}\}$$

$$\tilde{\mathscr{I}}_1 = [A_D X, 0, \bar{\delta}B_D Y_1, 0, (1 - \bar{\delta})B_D Y_2, 0, \tilde{\varepsilon}_1 I, \tilde{\varepsilon}_2 I]$$

$$\tilde{\mathscr{I}}_2 = [0, 0, BY_1, 0, -BY_2, 0, 0, 0], \quad \tilde{\Gamma}_{33} = \frac{\tilde{\Gamma}_{22}}{\bar{\delta}(1 - \bar{\delta})}$$

$$\tilde{\Gamma}_{41} = \begin{bmatrix} GX & 0 & 0 & 0 & 0 & 0 & 0 & 0 \\ H_1 X & 0 & 0 & 0 & 0 & 0 & 0 & 0 \\ 0 & 0 & \sqrt{\bar{\delta}}H_2 Y_1 & 0 & \tilde{\Lambda} & 0 & 0 & 0 \end{bmatrix}$$

$$\tilde{\Gamma}_{44} = diag\{-\tilde{\varepsilon}_1 I, -\tilde{\varepsilon}_2 I, -\tilde{\varepsilon}_2 I\}, \quad \tilde{\Lambda} = \sqrt{1 - \bar{\delta}}H_2 Y_2$$

with

$$\tilde{F}_{11} = AX + XA^T + \sum_{i=1}^{3} \frac{\tau_1 \tilde{Q}_i - R_i}{\tau_1}, \tilde{F}_{21} = \sum_{i=1}^{3} \frac{\tilde{R}_i}{\tau_1}$$

$$\tilde{F}_{22} = -\sum_{i=1}^{3} \frac{\tilde{R}_i}{\tau_1} - \frac{\tilde{R}_2 + \tilde{R}_3}{\tau_2 - \tau_1} - \tilde{Q}_1, \tilde{F}_{31} = \bar{\delta} Y_1^T B^T$$

$$\tilde{F}_{32} = \sum_{i=2}^{3} \frac{\tilde{R}_i - \tilde{U}_i}{\tau_2 - \tau_1}, \tilde{F}_{33} = \sum_{i=2}^{3} \frac{\tilde{U}_i^T + \tilde{U}_i - 2\tilde{R}_i}{\tau_2 - \tau_1}$$

$$\tilde{F}_{42} = \frac{\tilde{U}_2 + \tilde{U}_3}{\tau_2 - \tau_1}, \tilde{F}_{54} = \frac{\tilde{R}_3 - \tilde{U}_3}{\tau_3 - \tau_2}, \tilde{F}_{51} = \frac{Y_2^T B^T}{(1 - \bar{\delta})^{-1}}$$

$$\tilde{F}_{44} = -\tilde{Q}_2 - \frac{\tilde{R}_2 + \tilde{R}_3}{\tau_2 - \tau_1} - \frac{\tilde{R}_3}{\tau_3 - \tau_2}, \tilde{F}_{64} = \frac{\tilde{U}_3}{\tau_3 - \tau_2}$$

$$\tilde{F}_{55} = \frac{\tilde{U}_3^T + \tilde{U}_3 - 2\tilde{R}_3}{\tau_3 - \tau_2}, \tilde{F}_{66} = -\tilde{Q}_3 - \frac{\tilde{R}_3}{\tau_3 - \tau_2}$$

Proof Pre and postmultiply both sides of (5.27) with diag(X, X, X, X, X, X, ε_1^{-1}, ε_2^{-1}, I, I, I, I, I, ε_1^{-1}, ε_2^{-1}, ε_2^{-1}) and its transpose, and (5.28) with diag (X, X) and its transpose, respectively. Define $\tilde{\varepsilon}_i = \varepsilon_i^{-1}$ ($i = 1, 2$), $X = P^{-1}$, $XR_iX^T = \tilde{R}_i$, $XQ_iX^T = \tilde{Q}_i$ ($i = 1, 2, 3$), and $Y_i = K_iX^T$ ($i = 1, 2$). Then using Schur complement, we have (5.27). This completes the proof. $\qquad\qquad$ □

Notice that Theorem 5.2 cannot be directly implemented by the standard numerical software due to nonlinear terms $X\tilde{R}_i^{-1}X$. However, since $\tilde{R}_i > 0$ and $X > 0$, the conditions of $(\tilde{R}_i - X)\tilde{R}_i^{-1}(\tilde{R}_i - X) > 0$ are true and are equivalent to $-X\tilde{R}_i^{-1}X < \tilde{R}_i - 2X$. Then using $\tilde{R}_i - 2X$ to replace $-X\tilde{R}_i^{-1}X$ in Theorem 5.2, the original nonlinear matrix inequalities are transferred to LMIs, and the Matlab LMI Toolbox may be used to obtain the desired results. Moreover, the CCL approach [96] mentioned in Chaps. 3 and 4 also can be used to deal with nonlinear terms mentioned above.

5.3 Numerical Examples

In this section, two examples are introduced to show the effectiveness of the proposed method. One is two pendulums coupled by a spring (TPCS) [105–107], which is a classical test bed for the study of decentralized control of large unstable nonlinear systems; another is a quadruple-tank process [111, 112]. Here, both of the systems are to be extended to include the IP-based network in the communications.

Fig. 5.2 Two pendulums are coupled by a spring. © [2013] IEEE. Reprinted, with permission, from Ref. [105]

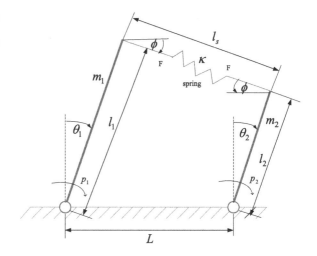

Example 1: As is shown in Fig. 5.2, two pendulums are coupled by a spring (TPCS), where each pendulum is treated as a subsystem and they are coupled by a spring between them [105–107].

The dynamic equations for the TPCS are given as

$$
\begin{aligned}
[m_1(l_1)^2/3]\ddot{\theta}_1 &= \pi_1 + m_1 g_1(l_1/2)\sin\theta_1 + l_1 F\cos(\theta_1 - \phi) \\
[m_1(l_2)^2/3]\ddot{\theta}_2 &= \pi_2 + m_2 g_1(l_2/2)\sin\theta_2 - l_2 F\cos(\theta_2 - \phi)
\end{aligned}
\tag{5.29}
$$

where $g_1 = 9.8\,\text{m/s}^2$ is the constant of gravity; θ_i is the angular displacement of pendulum i $(i = 1, 2)$; π_i is the torque input generated by the actuator for pendulum i $(i = 1, 2)$; F is the spring force; l_s is the spring length; ϕ is the angle of the spring to the earth; l_i is the length of pendulum i; m_i is the mass of pendulum; L is the distance of two pendulums; κ is the spring constant; and

$$
\begin{aligned}
F &= \kappa(l_s - [L^2 + (l_2 - l_1)^2]^{1/2}) \\
l_s &= \sqrt{(L + l_2\sin\theta_2 - l_1\sin\theta_1)^2 + (l_2\cos\theta_2 - l_1\cos\theta_1)^2} \\
\phi &= \tan^{-1}\frac{l_1\cos\theta_1 - l_2\cos\theta_2}{L + l_2\sin\theta_2 - l_1\sin\theta_1}
\end{aligned}
$$

The mass of each pendulum is uniformly distributed. The length of the spring is chosen so that $F = 0$ when $\theta_1 = \theta_2 = 0$, which implies that $(\theta_1\ \dot{\theta}_1\ \theta_2\ \dot{\theta}_2)^T = 0$ is an equilibrium of the system if $\pi_i = 0$. For simplicity, we assume that the mass of the spring is zero.

Define $x(t) = [\theta_1, \dot{\theta}_1, \theta_2, \dot{\theta}_2]^T$, $u(t) = [\pi_1(t), \pi_2(t)]^T$. Based on the above description, with the given $l_1 = 1\,\text{m}$, $l_2 = 0.8\,\text{m}$, $m_1 = 1\,\text{kg}$, $m_2 = 0.8\,\text{kg}$, $L = 1.2\,\text{m}$, $\kappa = 0.04\,\text{N/m}$, we obtain the state space model (5.1) with the nonlinear functions as described in (5.4) and (5.5) with the following parameters:

$$A = \begin{bmatrix} 0 & 1 & 0 & 0 \\ 13.7 & 0 & 0 & 0 \\ 0 & 0 & 0 & 1 \\ 0 & 0 & 17.2 & 0 \end{bmatrix}, g = \begin{bmatrix} 0 \\ 13.7(\sin\theta_1 - \theta_1) \\ 0 \\ 17.2(\sin\theta_2 - \theta_2) \end{bmatrix}$$

$$B = \begin{bmatrix} 0 & 0 \\ 3 & 0 \\ 0 & 0 \\ 0 & 5.9 \end{bmatrix}, f = \begin{bmatrix} 0 \\ 3F\cos(\theta_1 - \phi) \\ 0 \\ -4.68F\cos(\theta_2 - \phi) \end{bmatrix}$$

Note that the nonlinear functions in each subsystem S_i satisfies

$$|g_1(\sin\theta_i - \theta_i)| \le 0.45\,|\theta_i|\,,\,|\theta_i| \le \pi/6 \tag{5.30}$$

Then we have

$$
\begin{aligned}
&g^T(t, x(t))g(t, x(t)) \\
&\le (\frac{3}{2l_1}g_1)^2(\sin\theta_1 - \theta_1)^2 + (\frac{3}{2l_2}g_1)^2(\sin\theta_2 - \theta_2)^2 \\
&\le (\frac{3*0.45}{2l_1})^2\theta_1^2 + (\frac{3*0.45}{2l_2})^2\theta_2^2 = \|Gx(t)\|^2
\end{aligned}
\tag{5.31}
$$

where $G = diag\{0.675, 0, 0.844, 0\}$.

One can draw a three-dimensional figure of $|F|$ against $|\theta_1|$ and $|\theta_2|$, and obtain

$$|F| \le \kappa(1.49\,|\theta_1| + 0.18\,|\theta_2|) \tag{5.32}$$

From (5.32), it is clear

$$F^2 \le \kappa^2(2.738\theta_1^2 + 0.55\theta_2^2) \tag{5.33}$$

This leads to

$$
\begin{aligned}
&f^T(t, x(t))f(t, x(t)) \\
&= (\frac{3}{m_1 l_1}F)^2\cos^2(\theta_1 - \phi) + (\frac{3}{m_2 l_2}F)^2\cos^2(\theta_2 - \phi) \\
&\le (\frac{3}{m_1 l_1}F)^2 + (\frac{3}{m_2 l_2}F)^2 \le x^T(t)H_1^T H_1 x(t)
\end{aligned}
\tag{5.34}
$$

where $H_1 = diag\{0.368, 0, 0.165, 0\}$.

First, we consider a simple case, that is, all delays are assumed within the range of $[\tau_1, \tau_2]$ $(\delta(t) = 1)$. With the given lower delay bound $\tau_1 = 0.01\,s$, Table 5.1 gives the results of the two designs: one is based on the method given in [107] and the other is based on Theorem 5.2 of this chapter. From Table 5.1 one can see that the method given in this chapter is less conservative than that in [107]. Compared

Table 5.1 Upper delay bounds τ_2 and controller feedback gains for $\tau_1 = 0.01$ s and $\delta(t) = 1$

Method	τ_2	K
Yang et al. [107]	0.06	$\begin{pmatrix} -8.9190 & -2.4270 & -0.0034 & -0.0004 \\ -0.0018 & 0.0003 & -4.5521 & -1.2410 \end{pmatrix}$
Theorem 5.2	0.13	$\begin{pmatrix} -7.9824 & -2.0262 & 0 & 0 \\ 0 & 0 & -4.0850 & -1.0361 \end{pmatrix}$

with the centralized controller in [107], it is clear that the designed controller in this chapter is a decentralized controller since it only depends on its local subsystem state. Moreover, define the performance index $J = \sqrt{\int_0^T [x^T(s)x(s) + u^T(s)u(s)]ds}$. For a given simulation time $T = 5$ s, it is obtained J equals 141.3 and 150.3 for the centralized control in [107] and the decentralized control in this chapter, respectively. Furthermore, for a larger given spring constant $\kappa = 0.1$, we have $\tau_2 = 0.11$ s , J equals 140.7 and 164.4 in [107] and in this chapter, respectively. Compared with the cases of different spring constants, one can see the larger spring constant, the less allowable τ_2. Moreover, from the above results, it is observed that if all of the system's states are available at the same time, then smaller J may be expected in centralized control than that in decentralized control.

Second, considering that the communication delay is nonuniformly distributed. With the given $\bar{\delta} = 0.7$ and $\tau_1 = 0.01$ s, the results of the designs based on the method in this chapter are listed in Table 5.2. Comparing the results in Tables 5.1 and 5.2, one can see that the values of the upper bound τ_3 in Table 5.2 are larger than τ_2 in Table 5.1. This shows that if taking the communication characteristic into account, then less conservative results may be obtained.

For a given nonzero initial condition $x(t) = [-0.5, 0, 0.5, 0]^T$, Figs. 5.3 and 5.4 depict the nonuniform distribution IP-based network delay and the time responses of the system under consideration with 0.01 s $\leq \tau(t) < 0.28$ s and $Prob\{\tau(t) \in [0.01, 0.05)\} = 0.7$, $Prob\{\tau(t) \in [0.05, 0.28)\} = 0.3$.

Third, when there exists a communication network between the controller and the actuator, only the same controller gain can be used in (5.12). In this case, Table 5.3 also lists the obtained controller gain and upper delay bounds τ_3 with the given $\tau_1 = 0.01$ s and the different τ_2. Compared with the τ_3 listed in Tables 5.2 and 5.3, it is clear that the less conservative results can be obtained based on the distributed controller gain (5.12) than those based on the identical controller gain.

Table 5.2 Upper delay bounds τ_3 and controller feedback gains for the different values of τ_2 and given $\bar{\delta} = 0.7$

τ_2	τ_3	K_1	K_2
0.05	0.31	$\begin{pmatrix} -10.0265 & -2.9244 & 0 & 0 \\ 0 & 0 & -5.2126 & -1.1309 \end{pmatrix}$	$\begin{pmatrix} -5.2126 & -1.1309 & 0 & 0 \\ 0 & 0 & -2.6784 & -0.5780 \end{pmatrix}$
0.07	0.28	$\begin{pmatrix} -9.5750 & -2.8007 & 0 & 0 \\ 0 & 0 & -4.8707 & -1.4231 \end{pmatrix}$	$\begin{pmatrix} -5.6966 & -1.2015 & 0 & 0 \\ 0 & 0 & -2.9157 & -0.6166 \end{pmatrix}$
0.09	0.25	$\begin{pmatrix} -9.2688 & -2.6708 & 0 & 0 \\ 0 & 0 & -4.7368 & -1.3679 \end{pmatrix}$	$\begin{pmatrix} -5.9633 & -1.2461 & 0 & 0 \\ 0 & 0 & -3.0561 & -0.6360 \end{pmatrix}$

Fig. 5.3 IP-Based network
delay with
$0.01\,\mathrm{s} \leq \tau(t) < 0.28\,\mathrm{s}$

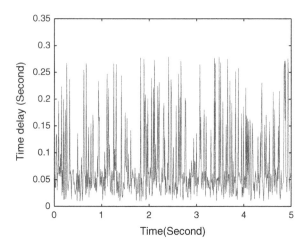

Time(Second)

Fig. 5.4 System's response
with $0.01\,\mathrm{s} \leq \tau(t) < 0.28\,\mathrm{s}$.
© [2013] IEEE. Reprinted,
with permission, from
Ref. [105]

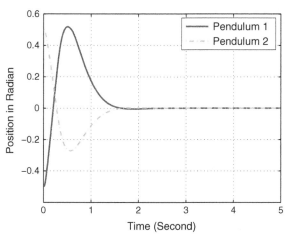

Time (Second)

Table 5.3 Upper delay bounds τ_3 and controller feedback gains for different values of τ_2 and given $\bar{\delta} = 0.7$

τ_2	τ_3	K
0.05	0.25	$\begin{pmatrix} -3.9481 & -1.0017 & 0 & 0 \\ 0 & 0 & -3.8056 & -0.9465 \end{pmatrix}$
0.07	0.24	$\begin{pmatrix} -7.7418 & -1.9654 & 0 & 0 \\ 0 & 0 & -3.9481 & -1.0017 \end{pmatrix}$
0.09	0.21	$\begin{pmatrix} -7.6746 & -1.9528 & 0 & 0 \\ 0 & 0 & -3.8979 & -0.9919 \end{pmatrix}$

Example 2: The quadruple-tank process consists of four interconnected water tanks and two pumps, the schematic diagram of the quadruple-tank process from [111] is shown in Fig. 5.5. The target is to control the level in the lower two tanks with two pumps through a communication network. The process inputs are v_1 and

Fig. 5.5 Schematic diagram of the quadruple-tank process. © [2013] IEEE. Reprinted, with permission, from Ref. [105]

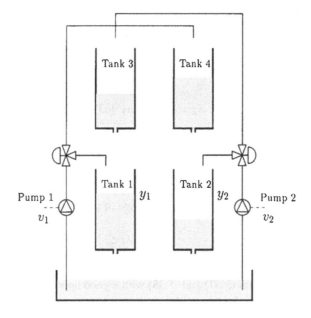

v_2 (input voltages to the pumps) and the outputs are y_1 and y_2 (voltages from level measurement devices). The nonlinear model equations are given as follows:

$$\begin{cases} \frac{dh_1}{dt} = \frac{a_1}{A_1}\sqrt{2gh_1} + \frac{a_3}{A_1}\sqrt{2gh_3} + \frac{\eta_1 k_1}{A_1}v_1 \\ \frac{dh_2}{dt} = \frac{a_2}{A_2}\sqrt{2gh_2} + \frac{a_4}{A_2}\sqrt{2gh_4} + \frac{\eta_2 k_2}{A_2}v_2 \\ \frac{dh_3}{dt} = \frac{a_3}{A_3}\sqrt{2gh_3} + \frac{(1-\eta_1)k_1}{A_3}v_1 \\ \frac{dh_4}{dt} = \frac{a_4}{A_4}\sqrt{2gh_4} + \frac{(1-\eta_2)k_2}{A_4}v_2 \\ y_i = k_c h_i, \quad i = 1, 2 \end{cases} \qquad (5.35)$$

where $A_1 = A_3 = 28\,\text{cm}^2$, $A_2 = A_4 = 32\,\text{cm}^2$ are cross-sections of the tanks, $a_1 = a_3 = 0.071\,\text{cm}^2$, $a_2 = a_4 = 0.057\,\text{cm}^2$ are cross-sections of the outlet holes, $g = 981\,\text{cm/s}^2$, $k_c = 0.5\,\text{V/cm}$. Suppose the operating range of the system is

$$0 \le h_i \le 20, \ i = 1, 2, \quad 0 \le h_i \le 6, \ i = 3, 4 \qquad (5.36)$$

For the minimum phase case, give the operating point

$$h_1 = 12.4\,\text{cm}, h_2 = 12.7\,\text{cm}, h_3 = 1.8\,\text{cm}, h_1 = 1.4\,\text{cm},$$
$$v_1 = v_2 = 3\,\text{V}, \eta_1 = 0.8, \eta_2 = 0.3, k_1 = 3.33\,\text{cm}^3/\text{V s},$$
$$k_2 = 3.35\,\text{cm}^3/\text{V s}.$$

Introducing $x_1 = h_1 - 12.4$, $x_2 = h_3 - 1.8$, $x_3 = h_2 - 12.7$, $x_4 = h_4 - 1.4$, $u_i = v_i - 3$, $i = 1, 2$, the system (5.35) can be evolved as:

$$\dot{x}_1 = 0.1123\sqrt{x_2 + 1.8} - 0.1123\sqrt{x_1 + 12.4}$$
$$+ 0.0951(u_1 + 3)$$
$$\dot{x}_2 = 0.0314(u_1 + 3) - 0.1123\sqrt{x_2 + 1.8}$$
$$\dot{x}_3 = 0.0789\sqrt{x_4 + 1.4} - 0.0789\sqrt{x_3 + 12.7} \qquad (5.37)$$
$$+ 0.0837(u_2 + 3)$$
$$\dot{x}_4 = 0.0208(u_2 + 3) - 0.1123\sqrt{x_1 + 1.4}$$

From (5.36), the operating range of the transformed system is

$$-12.4 \le x_1 \le 7.6, -1.8 \le x_2 \le 4.2$$
$$-12.7 \le x_3 \le 7.3, -1.4 \le x_4 \le 4.6 \qquad (5.38)$$

Based on (5.37) and (5.38) with a good approximation [112], we have

$$\dot{x}_1 = 0.0073x_1^2 - 0.017x_1 - 0.0047x_2^2$$
$$+ 0.049x_2 + 0.0951u_1$$
$$\dot{x}_2 = 0.0047x_2^2 - 0.049x_2 + 0.0314u_1$$
$$\dot{x}_3 = 0.0374x_4 + 0.0837u_2 - 0.00052x_3^2 \qquad (5.39)$$
$$- 0.011x_3 - 0.0035x_4^2$$
$$\dot{x}_4 = 0.0035x_4^2 - 0.0374x_4 + 0.0208u_2$$
$$y_1 = 0.5x_1, \; y_2 = 0.5x_3.$$

From (5.38) and (5.39), we have the model (5.1) with the following parameters:

$$A = \begin{bmatrix} -0.017 & 0.0492 & 0 & 0 \\ 0 & -0.0492 & 0 & 0 \\ 0 & 0 & -0.011 & 0.0374 \\ 0 & 0 & 0 & -0.0374 \end{bmatrix}$$

$$B = \begin{bmatrix} 0.0951 & 0 \\ 0 & 0 \\ 0 & 0.0837 \\ 0 & 0 \end{bmatrix}, \; H_2 = \begin{bmatrix} 0 & 0.0314 \\ 0.0208 & 0 \end{bmatrix}$$

$$G = \begin{bmatrix} 0.011 & 0 & 0 & 0 \\ 0 & 0.0036 & 0 & 0 \\ 0 & 0 & 0.011 & 0 \\ 0 & 0 & 0 & 0.004 \end{bmatrix}, \; H_1 = 0. \qquad (5.40)$$

Assume there is a communication network between the controllers and the actuators, and given $\tau_1 = 0.01$ s and $\tau_2 = 10$ s, $\tau_3 = 50$ s with $Prob\{\tau(t) \in [0.01, 10)\} = 0.7$. We obtain the following distributed controller gain

$$K = \begin{bmatrix} -0.0696 & -0.1365 & 0 & 0 \\ 0 & 0 & -0.1208 & -0.1769 \end{bmatrix} \tag{5.41}$$

Since there is a communication network between the controllers and the actuators, this is out of the scope of the proposed local decentralized PI control scheme in [112]. Moreover, B and G in (5.40) can also be written as

$$B = \begin{bmatrix} 0.0951 & 0 \\ 0 & 0.0314 \\ 0 & 0.0837 \\ 0.0208 & 0 \end{bmatrix}, H_1 = H_2 = 0 \tag{5.42}$$

Then we obtain the following controller gain with the same initial parameters as the above case.

$$K = \begin{bmatrix} -0.068139 & -0.14545 & 0.014992 & 0.027113 \\ 0.012094 & 0.044672 & -0.10936 & -0.16792 \end{bmatrix} \tag{5.43}$$

Compared with the decentralized controller based on (5.41), one can see that the controller based on (5.43) is a centralized controller since it depends on all of the system's states. From the given simulation time $T = 500$ s, it is obtained that J equals 92.3 and 85.7 for the controllers with the gains (5.41) and (5.43), respectively. Although in this case the centralized control outperforms the decentralized control performance under the network environment, the latter is more practical to a class of large-scale system controlled over a communication network [105].

5.4 Conclusion

The delay distribution-dependent decentralized control for a class of large-scale systems controlled over an IP-based network has been addressed, where the IP-based network delay has a probabilistic interval distribution characteristic. By coupling the decentralized networked control modeling and the delay distribution characteristic of IP-based communication networks in a unified framework, two delay distribution-dependent stability and stabilization conditions have been achieved. The designed decentralized controller depends both on the statistic distribution characteristic of the IP-based communication network and on the NCS's configuration. Simulation results have demonstrated the effectiveness of the theoretical results presented.

Part III
Necessary Communication Based Method for Control Design of NCSs

Chapter 6
H_∞ Filtering for NCSs with an Adaptive Event-Triggering Communication

In this chapter, an adaptive event-triggered communication scheme is used to save the limited network bandwidth, while preserving the desired H_∞ filter performance. Different from the continuous hardware-dependent event-triggered communication schemes in some existing ones, the proposed scheme requires that the sensor samples in a periodic manner, but whether the sampled data should be transmitted or not is determined by an adaptive event-triggered communication scheme. The idea of this chapter is to measure the output error between the current measurement instant and the latest transmitted measurement instant, the transmission should be executed only when an adaptively adjusted threshold is violated. Especially, a networked filtering-error system is modeled as a time-delay-dependent system to couple the proposed communication scheme and the filtering design in a unified framework.

This chapter is organized as follows. Section 6.1 proposes an adaptive event-triggered communication scheme to trigger the transmission event and an output-error-dependent model to couple the filter and the communication scheme in a unified framework. In Sect. 6.2, two sufficient criteria are derived for the analysis and design of a network filter system, respectively. Two examples are given in Sect. 6.3 to demonstrate the effectiveness of the proposed method. Section 6.4 concludes the chapter.

6.1 Communication and Model of Networked Filter

Consider the following discrete-time system:

$$\begin{cases} x(k+1) = Ax(k) + B\omega(k) \\ y(k) = Cx(k) + D\omega(k) \\ z(k) = Lx(k) \end{cases} \tag{6.1}$$

© Springer-Verlag Berlin Heidelberg 2015
C. Peng et al., *Communication and Control for Networked Complex Systems*,
DOI 10.1007/978-3-662-46813-5_6

where $x(k) \in \mathbb{R}^n$ is the state vector, $y(k) \in \mathbb{R}^m$ is the measured output, $z(k) \in \mathbb{R}^p$ is the signal to be estimated, $\omega(k) \in \mathbb{R}^q$ is assumed to be an arbitrary noise signal in $L_2[0, \infty)$. A, B, C, D, and L are constant matrices with appropriate dimensions.

For ease of exposition, the following assumptions are needed in this chapter:

- The sensors are time-triggered, and the event generators are event-triggered. The transmitted instant d_k is determined by the measured output $y(k)$. All transmitted packets are time-stamped.
- The filter is time-triggered and has a logic ZOH. The role of the logic ZOH is to accept a received packet only if the time stamp of the packet is greater than that of the packet currently stored in the ZOH.
- Network-induced delays from the sensor to the filter, and the computational and waiting delays are lumped together as $\hat{\tau}_{d_k}$, where $\hat{\tau}_{d_k} \in (0, \bar{\tau}]$, $\bar{\tau}$ is the upper bound of $\hat{\tau}_{d_k}$.

Throughout this chapter, we assume that the output signal $y(k)$ is transmitted over a communication network, and a networked H_∞ filter will be designed to estimate the $z(k)$ in (6.1). In what follows, we first propose an adaptive event-triggering communication scheme, then construct an output-error-dependent H_∞ filtering-error system modeling for the analysis and design of the studied system.

6.1.1 An Adaptive Event-Triggered Communication Scheme

In this subsection, an output-dependent adaptive event-triggered communication scheme is proposed to generate the transmission events by comparing the error between the measurements at the current sampling instant and the latest transmitted sampling instant, where the first event is generated at $d_0 = 0$, and any further transmission events are generated by the output-dependent threshold.

Similar to the adaptive event-triggered communication scheme described in Sect. 2.2.2, the next transmission instant determined by the adaptive event generator is expressed as

$$d_{k+1}h = d_k h + \min_{n \in \mathbb{N}}\{nh | e^T(i_k h)\Phi e(i_k h) > \sigma(t_k h)y^T(d_k h)\Phi y(d_k h)\} \quad (6.2)$$

where $\Phi > 0$ is a weighting matrix, and $i_k = d_k + i$, $(i = 0, 1, 2, \ldots, d_{k+1})$ is the current measurement instant; $e(i_k h)$ is the error between the current sampled data $y(i_k h)$ and the latest transmission data $y(d_k h)$, that is,

$$e(i_k h) \triangleq y(i_k h) - y(d_k h) \quad (6.3)$$

Moreover, $\sigma(t_k h)$ in (6.2) is determined by the following adaptive rule

$$\sigma(t_{k+1}h) = \max\{\sigma(t_k h)(1 - \frac{2\alpha}{\pi}\text{atan}[\beta(\|y(d_{k+1}h)\| - \|y(d_k h)\|)]), \sigma_m\} \quad (6.4)$$

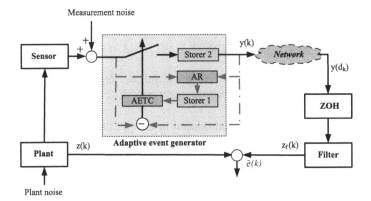

Fig. 6.1 A framework of networked filter with an adaptive event-triggered communication scheme

where atan$(\cdot)$ is the invert tangent function, $0 < \alpha$ and $0 < \beta$ are given constants to adjust the output of atan$(\cdot)$, σ_m is the given lower bound of $\sigma(t_k h)$, $\sigma(0) = \sigma_m$.

From (6.2) and (6.4), one can see that the transmission events are dependent both on the error $e(i_k h)$ and the latest transmitted $y(d_k h)$, and on the adjustable threshold $\sigma(t_k h)$.

The structure of proposed adaptive event-triggered communication scheme is shown in Fig. 6.1, where adaptive rule (AR) is given in (2.19); the role of Store 1 and 2 is mentioned in Sect. 2.2.2; the adaptive event-triggered communication scheme (AETC) is located downstream of the sensor; the sensor measures with a constant sampling period. Whether or not the sampled data need to be transmitted is determined by the predetermined communication scheme. In other words, the adaptive event generator is running in a periodic manner.

6.1.2 Modeling of Networked Filtering-Error Systems

Suppose that system (6.1) is asymptotically stable. In this case, a networked full-order filter is designed for the estimation of $z(t)$ with the following state-space realization:

$$\begin{cases} x_f(k+1) = A_f x_f(k) + B_f \hat{y}(k) \\ z_f(k) = C_f x_f(k) + D_f \hat{y}(k) \end{cases} \tag{6.5}$$

where $A_f \in \mathbb{R}^{n \times n}$, $B_f \in \mathbb{R}^{n \times m}$, $C_f \in \mathbb{R}^{p \times n}$ and $D_f \in \mathbb{R}^{p \times m}$ are the filter parameters to be determined, $x_f(0) = 0$.

Different from the traditional filtering problem [113], since the data of the measured output $y(k)$ in (6.1) is sampled and transmitted through a communication network, $\hat{y}(k)$ in (6.5) does not equal to $y(k)$ in (6.1). Considering the nonideal network QoS, (6.5) can be expressed as

$$\begin{cases} x_f(k+1) = A_f x_f(k) + B_f y(d_k) \\ z_f(k) = C_f x_f(k) + D_f y(d_k) \\ k \in \Omega \triangleq [d_k + \tau_{d_k}, d_{k+1} + \tau_{d_{k+1}}), \quad k = 1, 2, 3, \dots \end{cases} \quad (6.6)$$

where d_k $(k = 1, 2, 3, \dots)$ are some integrals, and $\{d_1, d_2, d_3, \dots\} \subset \{1, 2, 3, \dots\}$; τ_{d_k} are the integer multiples of the measurement period; Ω is the holding interval of the logic ZOH.

By making use of the method introduced in Chap. 2, the holding interval Ω in (6.6) of ZOH is also divided into sampling-interval-like subsets $\Omega_i \triangleq [i_k + \tau_{d_k+i},$ $i_k + 1 + \tau_{d_k+i+1})$, i.e., $\Omega = \cup \Omega_i$, where $i = 0, 1, \dots, d_{k+1} - d_k - 1$. If i equal to 0 or $d_{k+1} - d_k - 1$, τ_{d_k+i} mean the practical communication delay, otherwise, τ_{d_k+i} can be chosen for keeping the condition of $\tau_{d_k+i} \leq 1 + \tau_{d_k+i+1}$ in Ω_i. Moreover, it is seen that when $t \in \Omega_i$, the output of the ZOH is invariant until a new transmission event being triggered [114].

Define

$$\begin{cases} e(i_k) \triangleq y(i_k) - y(d_k) \\ \eta(k) \triangleq k - i_k, \ k \in \Omega_i \end{cases} \quad (6.7)$$

Then, the input of the filter $y(d_k)$ described in (6.6) can be represented as

$$y(d_k) = y(k - \eta(k)) - e(i_k), \quad k \in \Omega_i \quad (6.8)$$

Define $\tilde{x}(k) \triangleq [x^T(k) \ x_f^T(k)]^T$, $\tilde{e}(k) \triangleq z(k) - z_f(k)$, $\tilde{\omega}(k) \triangleq [\omega^T(k)$ $\omega^T(k - \eta(k))]^T$. From (6.1), (6.6) and (6.8), the following time-delay dependent filtering-error system can be obtained:

$$\tilde{x}(k+1) = \tilde{A}(k) + \tilde{B}H\tilde{x}(k - \eta(k)) + \tilde{C}\tilde{\omega}(k) + \tilde{D}e(i_k) \quad (6.9a)$$

$$\tilde{e}(k) = \tilde{L}\tilde{x}(k) + \tilde{E}H\tilde{x}(k - \eta(k)) + \tilde{F}\tilde{\omega}(k) + D_f e(i_k) \quad (6.9b)$$

where

$$\tilde{A} = \begin{bmatrix} A & 0 \\ 0 & A_f \end{bmatrix}, \tilde{B} = \begin{bmatrix} 0 \\ B_f C \end{bmatrix}, \tilde{C} = \begin{bmatrix} B & 0 \\ 0 & B_f D \end{bmatrix},$$

$$\tilde{D} = \begin{bmatrix} 0 \\ -B_f \end{bmatrix}, H = \begin{bmatrix} I & 0 \end{bmatrix}, \tilde{L} = \begin{bmatrix} L & -C_f \end{bmatrix},$$

$$\tilde{E} = -D_f C, \tilde{F} = \begin{bmatrix} 0 & -D_f D \end{bmatrix}.$$

Notice that the output error $e(i_k)$ is included in the time-delay dependent filtering-error model (6.9), which is convenient for us to have the H_∞ filter analysis and design, while using the propose event-triggered communication scheme to determine whether or not the current measurement should be transmitted. Moreover, from the definition of $\eta(k)$ in (6.7), one can see that $\eta(k)$ is a function satisfying

$$0 < \eta_1 \leq \eta(k) \leq \eta_3, t \in \Omega_i \tag{6.10}$$

where $\eta_1 \triangleq min_k\{\tau_{d_k+i}\}$, $\eta_3 \triangleq max_k\{1 + \tau_{d_k+i+1}\} = h + \bar{\tau}$.

For convenience, the following definition is introduced.

Definition 6.1 The system (6.9) is asymptotically stable with an H_∞ norm bound γ, if
(i) The system (6.9) with $\tilde{\omega}(k) = 0$ is asymptotically stable;
(ii) The system (6.9) has a prescribed level γ of H_∞ noise attenuation, i.e., under the assumption of zero initial condition, the condition of $\|\tilde{e}(k)\|_2 < \gamma \|\tilde{\omega}(k)\|_2$ is satisfied for any nonzero $\tilde{\omega}(k) \in L_2[0, \infty)$.

6.2 H_∞ Filtering Analysis and Design

In this section, an H_∞ filtering analysis result and an H_∞ filtering design result for system (6.9) are presented.

Theorem 6.1 *For some given constants* $0 \leq \eta_1 \leq \eta_3$ *and* γ, *under the given communication scheme* (6.2), *the filtering-error system* (6.9) *is asymptotically stable with an* H_∞ *norm bound* γ, *if there exist real matrices* $P > 0$, $Q_l > 0$, $R_l > 0$ *($l = 1, 2, 3$) and matrices* U_j *($j = 2, 3$) with appropriate dimensions such that for* $i = 1, 2$ *and* $j = 2, 3$

$$\Pi^i = \begin{bmatrix} \Pi_{11}^i & \Pi_{12} & \Pi_{13} \\ * & \Pi_{22} & 0 \\ * & * & \Pi_{33} \end{bmatrix} < 0 \tag{6.11}$$

$$\mathbb{U}_j = \begin{bmatrix} R_j & U_j^T \\ * & R_j \end{bmatrix} \geq 0 \tag{6.12}$$

where

$$\Pi_{11}^i = \begin{bmatrix} \Gamma_{11} & H^T R_1 & P\tilde{B} & 0 & 0 & P\tilde{D} & P\tilde{C} \\ * & \Gamma_{22} & \Gamma_{23}^i & \Gamma_{24}^i & 0 & 0 & 0 \\ * & * & \Gamma_{33}^i & \Gamma_{34}^i & \Gamma_{35}^i & \Gamma_{36} & 0 \\ * & * & * & \Gamma_{44} & \Gamma_{45}^i & 0 & 0 \\ * & * & * & * & \Gamma_{55} & 0 & 0 \\ * & * & * & * & * & \Gamma_{66} & \Gamma_{67} \\ * & * & * & * & * & * & -\gamma^2 I \end{bmatrix},$$

$$\Pi_{12} = \begin{bmatrix} \tilde{A} - I & 0 & \tilde{B} & 0 & 0 & \tilde{D} & \tilde{C} \end{bmatrix}^T \begin{bmatrix} P & H^T & H^T & H^T \end{bmatrix}$$

$$\Pi_{13} = \begin{bmatrix} \tilde{L} & 0 & \tilde{E} & 0 & 0 & D_f & \tilde{F} \\ 0 & 0 & \delta_m \Phi C & 0 & 0 & 0 & \delta_m \Phi \begin{bmatrix} 0 & D \end{bmatrix} \end{bmatrix}^T$$

$$\Pi_{22} = -diag\{P, \frac{R_1^{-1}}{\eta_1^2}, \frac{R_2^{-1}}{(\eta_2 - \eta_1)^2}, \frac{R_3^{-1}}{(\eta_3 - \eta_2)^2}\},$$

$$\Pi_{33} = -diag\{I, \Phi\},$$

with

$$\Gamma_{11} = P(\tilde{A} - I) + (\tilde{A}^T - I)P + H^T(\sum_{i=1}^{3} Q_i - R_1)H$$

$$\Gamma_{22} = -Q_1 - R_1 - R_2, \Gamma_{23}^i = (2 - i)(R_2 - U_2^T)$$

$$\Gamma_{24}^i = (2 - i)U_2^T - (1 - i)R_2, \Gamma_{33}^i = U_{i+1} + U_{i+1}^T - 2R_{i+1}$$

$$\Gamma_{34}^i = (2 - i)(R_2 - U_2^T) + (1 - i)(U_3 - R_3)$$

$$\Gamma_{35}^i = (1 - i)(U_3^T - R_3), \Gamma_{36} = -\delta_m C^T \Phi$$

$$\Gamma_{44} = -Q_2 - R_3 - R_2, \Gamma_{45}^i = (2 - i)R_3 - (1 - i)U_3^T$$

$$\Gamma_{55} = -Q_3 - R_3, \Gamma_{66} = -(1 - \delta_m)\Phi, \Gamma_{67} = -\delta_m \Phi[0\ D]$$

Proof Construct a Lyapunov–Krasovskii functional candidate as

$$V(k) = \tilde{x}^T(k)P\tilde{x}(k) + \sum_{i=1}^{3} \sum_{j=k-\eta_i}^{k-1} \tilde{x}^T(j)H^T Q_i H\tilde{x}(j)$$

$$+ \sum_{l=1}^{3} \sum_{i=-\eta_l}^{-\eta_{l-1}-1} \sum_{\theta=k+i}^{k-1} (\eta_l - \eta_{l-1})\mathfrak{y}^T(\theta)H^T R_l H\mathfrak{y}(\theta) \qquad (6.13)$$

where $\eta_0 = 0$; $\mathfrak{y}(\theta) = \tilde{x}(\theta+1) - \tilde{x}(\theta)$; $\eta_2 := \frac{\eta_1+\eta_3}{2}$ if $\frac{\eta_3-\eta_1}{2}$ is an integer, else $\eta_2 := \frac{\eta_3+\eta_1+1}{2}$.

Defining $\Delta V(k) = V(k+1) - V(k)$ yields

$$\Delta V(k) = 2\tilde{x}^T(k)P\mathfrak{y}(k) + \mathfrak{y}^T(k)P\mathfrak{y}(k) + \sum_{i=1}^{3} \tilde{x}^T(k)H^T Q_i H\tilde{x}(k)$$

$$- \sum_{i=1}^{3} \tilde{x}^T(k - \eta_i)H^T Q_i H\tilde{x}(k - \eta_i) + \sum_{l=1}^{3} \mathfrak{y}^T(k)(\eta_l - \eta_{l-1})^2 H^T R_l H\mathfrak{y}(k)$$

$$- \sum_{l=1}^{3} \sum_{\theta=k-\eta_l}^{k-\eta_{l-1}-1} (\eta_l - \eta_{l-1})\mathfrak{y}^T(\theta)H^T R_l H\mathfrak{y}(\theta) + e^T(i_k)\Phi e(i_k) - e^T(i_k)\Phi e(i_k)$$

$$(6.14)$$

For notational simplicity, define $\xi^T(k) = [\underbrace{\tilde{x}^T(k)}_{\lambda_1}, \underbrace{\tilde{x}^T(k - \eta_1)H^T}_{\lambda_2}, \underbrace{\tilde{x}^T(k - \eta(k))H^T}_{\lambda_3},$

$\underbrace{\tilde{x}^T(k - \eta_2)H^T}_{\lambda_4}, \underbrace{\tilde{x}^T(k - \eta_3)H^T}_{\lambda_5}, e^T(i_kh), \tilde{\omega}^T(k)]^T.$

Now, we prove our result for $\eta_3 \geq \eta(t) \geq \eta_2$ and $\eta_2 > \eta(t) \geq \eta_1$, respectively.

Case I: $\eta_2 > \eta(t) \geq \eta_1$. For any $\mathbb{U}_2 = \begin{bmatrix} R_2 & U_2^T \\ U_2 & R_2 \end{bmatrix} \geq 0$, using discrete Lemmas 2.2 and 2.7 to deal with sum items in (6.14), we have

$$- \eta_1 \sum_{j=k-\eta_1}^{k-1} \mathfrak{y}^T(j)H^T R_1 H \mathfrak{y}(j) \leq -[\lambda_1 - \lambda_2]^T R_1[\lambda_1 - \lambda_2] \tag{6.15}$$

$$- (\eta_3 - \eta_2) \sum_{j=k-\eta_3}^{k-\eta_2-1} \mathfrak{y}^T(j)H^T R_3 H \mathfrak{y}(j) \leq -[\lambda_4 - \lambda_5]^T R_3[\lambda_4 - \lambda_5] \tag{6.16}$$

$$- (\eta_2 - \eta_1) \sum_{j=k-\eta_2}^{k-\eta_1-1} \mathfrak{y}^T(j)H^T R_2 H \mathfrak{y}(j) \leq - \begin{bmatrix} \lambda_2 - \lambda_3 \\ \lambda_3 - \lambda_4 \end{bmatrix}^T \mathbb{U}_2 \begin{bmatrix} \lambda_2 - \lambda_3 \\ \lambda_3 - \lambda_4 \end{bmatrix} \tag{6.17}$$

Moreover, for $k \in \Omega_i$, the proposed event-triggering communication scheme (6.2) ensures that

$$e^T(i_k)\Phi e(i_k) < \delta_m y^T(d_k)\Phi y(d_k)$$
$$= \delta_m[e(i_k) - y(k - \eta(k))]^T \Phi[e(i_k) - y(k - \eta(k))]$$
$$= \delta_m e^T(i_k)\Phi e(i_k) - 2\delta_m[Cx(k - \eta(k))$$
$$+ [0 \ D]\tilde{\omega}(k)]^T \Phi e(i_k) + \delta_m[Cx(k - \eta(k))$$
$$+ [0 \ D]\tilde{\omega}(k)]^T \Phi[Cx(k - \eta(k)) + [0 \ D]\tilde{\omega}(k)] \tag{6.18}$$

From (6.14)–(6.18), $\Delta V(k)$ in (6.14) is evolved as

$$\Delta V(k) \leq \xi^T(k)[\Pi_{11}^1 - \Pi_{12}\Pi_{22}^{-1}\Pi_{21}^T]\xi(k) - \tilde{e}^T(k)\tilde{e}(k) + \gamma^2\tilde{\omega}^T(k)\tilde{\omega}(k) \tag{6.19}$$

Case II: $\eta_3 \geq \tau(t) \geq \eta_2$. Keeping (6.15) and utilizing Lemma 2.7 and discrete Lemma 2.2 to deal with accumulative items of (6.16) and (6.17), respectively, that is, for any $\mathbb{U}_3 = \begin{bmatrix} R_3 & U_3^T \\ U_3 & R_3 \end{bmatrix} \geq 0$,

$$- (\eta_3 - \eta_2) \sum_{j=k-\eta_3}^{k-\eta_2-1} \mathfrak{y}^T(j)H^T R_3 H \mathfrak{y}(j) \leq - \begin{bmatrix} \lambda_4 - \lambda_3 \\ \lambda_3 - \lambda_5 \end{bmatrix}^T \mathbb{U}_3 \begin{bmatrix} \lambda_4 - \lambda_3 \\ \lambda_3 - \lambda_5 \end{bmatrix} \tag{6.20}$$

$$- (\eta_2 - \eta_1) \sum_{j=k-\eta_2}^{k-\eta_1-1} \mathfrak{y}^T(j)H^T R_2 H \mathfrak{y}(j) \leq -[\lambda_2 - \lambda_4]^T R_2[\lambda_2 - \lambda_4] \tag{6.21}$$

Then, from (6.14), (6.15), (6.20) and (6.21), $\Delta V(k)$ in (6.14) is evolved as

$$\Delta V(k) \leq \xi^T(k)[\Pi_{11}^2 - \Pi_{12}\Pi_{22}^{-1}\Pi_{21}^T]\xi(k) - \tilde{e}^T(k)\tilde{e}(k) + \gamma^2\tilde{\omega}^T(k)\tilde{\omega}(k) \quad (6.22)$$

First, we show that the system (6.9) with $\tilde{\omega}(k) = 0$ is asymptotically stable. From Schur complement, the conditions of $\Pi^i < 0$ $(i = 1, 2)$ in (6.11) guarantee that $\Pi_{11}^i - \Pi_{21}^T\Pi_{22}^{-1}\Pi_{21} < 0$ in (6.19) and (6.22). Then, it is readily obtained that

$$\Delta V(k) \leq \xi^T(k)[\Pi_{11}^i - \Pi_{12}\Pi_{22}^{-1}\Pi_{21}^T]\xi(k) - \tilde{e}^T(k)\tilde{e}(k) < 0, \quad (6.23)$$

which guarantees the asymptotic stability of the system (6.9) with $\tilde{\omega}(k) = 0$.

Next, assuming under the zero initial condition, we show that the condition of $\|\tilde{e}(k)\|_2 < \gamma \|\tilde{\omega}(k)\|_2$ is true for all nonzero $\tilde{\omega}(k) \in L_2[0, \infty)$. Again, the conditions of $\Pi^i < 0$ $(i = 1, 2)$ in (6.11) guarantee that $\Pi_{11}^i - \Pi_{21}^T\Pi_{22}^{-1}\Pi_{21} < 0$ in (6.19) and (6.22). Then, it is clear that

$$\Delta V(k) < \gamma^2\tilde{\omega}^T(k)\tilde{\omega}(k) - \tilde{e}^T(k)\tilde{e}(k) \quad (6.24)$$

Summing both sides of (6.24) from 0 to ∞, we have

$$\sum_{k=0}^{\infty} \tilde{e}^T(k)\tilde{e}(k) < \sum_{k=0}^{\infty} \gamma^2\tilde{\omega}^T(k)\tilde{\omega}(k) + V(0) - V(\infty) \quad (6.25)$$

Letting $k \to \infty$ and under the zero initial condition, from (6.25) we have

$$\sum_{k=0}^{\infty} \tilde{e}^T(k)\tilde{e}(k) < \sum_{k=0}^{\infty} \gamma^2\tilde{\omega}^T(k)\tilde{\omega}(k) \quad (6.26)$$

That is, $\|\tilde{e}(k)\|_2 < \gamma \|\tilde{\omega}(k)\|_2$ is true for all nonzero $\tilde{\omega}(k) \in L_2[0, \infty)$. Therefore, by Definition 6.1, the result is established. This completes the proof. □

If we consider the following system as [115], by making use of the similar proof technique as in the derivation of Theorem 6.1, we have Corollary 6.1.

$$\begin{cases} x(k+1) = Ax(k) + Bu(k) \\ u(k) = Kx(k - \eta(k)) \end{cases} \quad (6.27)$$

Corollary 6.1 *For given scalars η_1, η_3 and matrix K, the system (6.27) is asymptotically stable if there exist matrices $P > 0$, $Q_i > 0$, $R_i > 0$ $(i = 1, 2, 3)$ and U_j $(j = 1, 2)$ with appropriate dimensions, such that the following LMIs hold for $l = 1, 2$*

$$\begin{bmatrix} \Xi_{11}^l & * \\ \Xi_{21} & \Xi_{22} \end{bmatrix} < 0$$

$$\begin{bmatrix} R_{l+1} & * \\ U_l & R_{l+1} \end{bmatrix} > 0$$

where

$$\Xi_{11}^1 = \begin{bmatrix} \Phi_{11} & * & * & * & * \\ R_1 & \Phi_{22} & * & * & * \\ 0 & U_1 & \Phi_{33} & * & * \\ 0 & 0 & R_3 & -Q_3 - R_3 & * \\ K^T B^T P & R_2 - U_1 & R_2 - U_1^T & 0 & \Phi_{55} \end{bmatrix}$$

$$\Xi_{11}^2 = \begin{bmatrix} \Phi_{11} & * & * & * & * \\ R_1 & \Phi_{22} & * & * & * \\ 0 & R_2 & \Phi_{33} & * & * \\ 0 & 0 & U_2 & -Q_3 - R_3 & * \\ K^T B^T P & 0 & R_3 - U_2 & R_3 - U_2^T & \Phi_{55} \end{bmatrix}$$

$$\Xi_{21} = [F^T P \quad \eta_1 F^T R_1 \quad (\eta_2 - \eta_1)F^T R_2 \quad (\eta_3 - \eta_2)F^T R_3]^T,$$
$$\Xi_{22} = diag\{P, \ R_1, \ R_2, \ R_3\}, F = [A - I \ 0 \ 0 \ 0 \ BK]$$
$$\Phi_{11} = PA + A^T P - 2P + Q_1 + Q_2 + Q_3 - R_1, \Phi_{22} = -Q_1 - R_1 - R_2$$
$$\Phi_{33} = -Q_2 - R_2 - R_3, \Phi_{55} = -2R_2 + U_1 + U_1^T$$

In the following, we are seeking to design the H_∞ filter based on Theorem 6.1.

Theorem 6.2 *For some given constants $0 \leq \eta_1 \leq \eta_3$ and γ, under the given adaptive communication scheme (6.2), the filtering-error system (6.9) is asymptotically stable with an H_∞ norm bound γ, and the filter parameter matrices of the filter (6.6) are given as*

$$\begin{bmatrix} A_f & B_f \\ C_f & D_f \end{bmatrix} = \begin{bmatrix} \bar{A}_f \bar{P}_3^{-1} & \bar{B}_f \\ \bar{C}_f \bar{P}_3^{-1} & \bar{D}_f \end{bmatrix}, \tag{6.28}$$

if there exist real matrices $P_1 > 0$, $\bar{P}_3 > 0$, $Q_l > 0$, $R_l > 0$ ($l = 1, 2, 3$) and matrices U_j ($j = 2, 3$) and M_l with appropriate dimensions such that for $i = 1, 2$ and $j = 2, 3$

$$\tilde{\Pi}^i = \begin{bmatrix} \tilde{\Pi}_{11}^i & \tilde{\Pi}_{12} & \tilde{\Pi}_{13} \\ * & \tilde{\Pi}_{22} & 0 \\ * & * & \tilde{\Pi}_{33} \end{bmatrix} < 0 \tag{6.29}$$

$$\mathbb{U}_j = \begin{bmatrix} R_j & U_j^T \\ * & R_j \end{bmatrix} \geq 0 \tag{6.30}$$

where

$$\tilde{\Pi}_{11}^i = \begin{bmatrix} \Psi_{11} & \Psi_{12} \\ * & \Psi_{22} \end{bmatrix},$$

$$\Psi_{11} = \begin{bmatrix} \Lambda_{11} & \Lambda_{12} & R_1 & \bar{B}_f C & 0 & 0 & -\bar{B}_f \\ * & \Lambda_{22} & 0 & \bar{B}_f C & 0 & 0 & -\bar{B}_f \\ * & * & \Gamma_{22} & \Gamma_{23}^i & \Gamma_{24}^i & 0 & \Lambda_{26} \\ * & * & * & \Gamma_{33}^i & \Gamma_{34}^i & \Gamma_{35}^i & 0 \\ * & * & * & * & \Gamma_{44} & \Gamma_{45}^i & 0 \\ * & * & * & * & * & \Gamma_{55} & 0 \\ * & * & * & * & * & * & \Lambda_{66} \end{bmatrix}$$

$$\Psi_{12} = \begin{bmatrix} P_1 B & \bar{B}_f D \\ \bar{P}_3 B & \bar{B}_f D \\ 0 & 0 \\ 0 & 0 \\ 0 & 0 \\ 0 & 0 \\ 0 & -\delta_m \Phi D \end{bmatrix}, \Psi_{22} = \begin{bmatrix} -\gamma^2 I & 0 \\ * & -\gamma^2 I \end{bmatrix}$$

$$\tilde{\Pi}_{12} = \begin{bmatrix} (A^T - I) P_1 & (A^T - I) \bar{P}_3 \\ \bar{A}_f^T - \bar{P}_3^T & \bar{A}_f^T - \bar{P}_3^T \\ 0 & 0 \\ C^T \bar{B}_f^T & C^T \bar{B}_f^T \\ 0 & 0 \\ 0 & 0 \\ -\bar{B}_f^T & -\bar{B}_f^T \\ B^T P_1^T & B^T \bar{P}_3^T \\ D^T \bar{B}_f^T & D^T \bar{B}_f^T \end{bmatrix}, \tilde{\Pi}_{13} = \begin{bmatrix} \Psi_{31}^T \\ \Psi_{31}^T \end{bmatrix},$$

$$\Psi_{31} = \begin{bmatrix} M_1(A - I) & 0 & 0 & 0 & 0 & 0 & 0 \\ M_2(A - I) & 0 & 0 & 0 & 0 & 0 & 0 \\ M_3(A - I) & 0 & 0 & 0 & 0 & 0 & 0 \\ L & -\bar{C}_f & 0 & -D_f C & 0 & 0 & D_f \\ 0 & 0 & 0 & \delta_m \Phi C & 0 & 0 & 0 \end{bmatrix}^T$$

$$\Psi_{32} = \begin{bmatrix} M_1 B & 0 \\ M_2 B & 0 \\ M_3 B & 0 \\ 0 & -D_f D \\ 0 & \delta_m \Phi D \end{bmatrix}^T, \tilde{\Pi}_{22} = -\begin{bmatrix} P_1 & \bar{P}_3 \\ \bar{P}_3 & \bar{P}_3 \end{bmatrix}$$

$$\tilde{\Pi}_{33} = -diag\{\eta_1^{-2} M_1 R_1^{-1} M_1^T, (\eta_2 - \eta_1)^{-2} M_2 R_2^{-1} M_2^T,$$
$$(\eta_3 - \eta_2)^{-2} M_3 R_3^{-1} M_3^T, I, \Phi\}.$$

with

$$
\Lambda_{11} = P_1(A - I) + (A - I)^T P_1 + \sum_{i=1}^{3} Q_i - R_1
$$

$$
\Lambda_{12} = \bar{A}_f - \bar{P}_3 + (A - I)^T \bar{P}_3, \Lambda_{26} = -\delta_m C^T \Phi
$$

$$
\Lambda_{22} = \bar{A}_f + \bar{A}_f^T - 2\bar{P}_3, \Lambda_{66} = -(1 - \delta_m)\Phi
$$

Proof Define

$$
P = \begin{bmatrix} P_1 & P_2^T \\ P_2 & P_3 \end{bmatrix}, J = \begin{bmatrix} I & 0 \\ 0 & P_2^T P_3^{-1} \end{bmatrix} \tag{6.31}
$$

where $P_1, P_3 \in \mathbb{R}^{n \times n}$. From $P > 0$, we obtain that $P_3 + P_3^T > 0$ and P_3 is invertible. Multiply both side of (6.11) with $\Gamma = diag\{J, I, I, I, I, I, I, J, M_1, M_2, M_3, I, I\}$ and its transpose, where M_i ($i = 1, 2, 3$) are matrices with appropriate dimensions, and set

$$
\bar{A}_f = P_2^T A_f P_3^{-1} P_2, \bar{B}_f = P_2^T B_f,
$$

$$
\bar{C}_f = C_f P_3^{-1} P_2, \bar{D}_f = D_f, \bar{P}_3 = P_2^T P_3^{-1} P_2. \tag{6.32}
$$

Notice that

$$
JP(\tilde{A} - I)J^T = \begin{bmatrix} P_1(A - I) & \bar{A}_f - \bar{P}_3 \\ \bar{P}_3(A - I) & \bar{A}_f - \bar{P}_3 \end{bmatrix}
$$

$$
JH^T(\sum_{i=1}^{3} Q_i - R_1)HJ^T = \begin{bmatrix} \sum_{i=1}^{3} Q_i - R_1 & 0 \\ 0 & 0 \end{bmatrix}
$$

$$
JH^T R_1 = \begin{bmatrix} R_1 \\ 0 \end{bmatrix}, JP\tilde{B} = \begin{bmatrix} \bar{B}_f C \\ \bar{B}_f C \end{bmatrix}
$$

$$
JP\tilde{D} = \begin{bmatrix} -\bar{B}_f \\ -\bar{B}_f \end{bmatrix}, JP\tilde{C} = \begin{bmatrix} P_1 B & \bar{B}_f D \\ \bar{P}_3 B & \bar{B}_f D \end{bmatrix},
$$

then, (6.29) can be obtained from (6.11).

In the following, we seek to obtain the other filter parameters A_f, B_f, C_f and D_f. First, if (6.29) is feasible, we have $\bar{P}_3 > 0$, which implies that P_2 is invertible from $\bar{P}_3 = P_2^T P_3^{-1} P_2$. Then the filter parameters can be written as

$$
A_f = P_2^{-T} \bar{A}_f \bar{P}_3^{-1} P_2^T, B_f = P_2^{-T} \bar{B}_f, C_f = \bar{C}_f \bar{P}_3^{-1} P_2^T \tag{6.33}
$$

Since the following systems are algebraically equivalent [116]

$$\begin{bmatrix} A_f & B_f \\ C_f & D_f \end{bmatrix} = \begin{bmatrix} P_2^{-T} \bar{A}_f \bar{P}_3^{-1} P_2^T & P_2^{-T} \bar{B}_f \\ \bar{C}_f \bar{P}_3^{-1} P_2^T & \bar{D}_f \end{bmatrix} \Longleftrightarrow \begin{bmatrix} \bar{A}_f \bar{P}_3^{-1} & \bar{B}_f \\ \bar{C}_f \bar{P}_3^{-1} & \bar{D}_f \end{bmatrix} \quad (6.34)$$

thus, a state-space realization as (6.28) of the desired filter is readily obtained from (6.34). This completes the proof. $\square$

Notice that the matrix inequality (6.29) is a non-convex inequality due to the existence of the items $M_i R_i^{-1} M_i^T$ ($i = 1, 2, 3$) in $\tilde{\Pi}_{33}$. Generally, there are two methods that can be utilized to solve such non-convex minimization problem by Matlab LMI Toolbox: one is cone complementarity numerical approach (CCL) [96]; another is simpler linear approach, e.g., for any matrices M_i ($i = 1, 2, 3$) with appropriate dimensions, it is readily derived that

$$- M_i R_i^{-1} M_i^T \leq R_i - M_i - M_i^T \ (i = 1, 2, 3). \quad (6.35)$$

Then, using $R_i - M_i - M_i^T$ to replace $-M_i^T R_i^{-1} M_i$ in (6.29), the original non-convex minimization problem is transferred to convex minimization problem. Although the CCL results are slightly less conservative than those based on (6.35), it needs more auxiliary variables in solving LMIs, therefore, the latter method is employed in this chapter.

6.3 Illustrate Examples

In this section, two examples are provided to show the effectiveness of the proposed method. Example 1 is used to show that improved results can be obtained by making use of Lemma 2.7 in the derivation of the result; Example 2 is used to show that with the proposed adaptive event-triggered communication scheme, the same H_∞ performance can be obtained as those obtained from the time-triggered communication scheme, while using less communication bandwidth.

Example 1: Consider the system (6.27) with the following parameters

$$A = \begin{bmatrix} 0.8 & 0 \\ 0.05 & 0.9 \end{bmatrix}, BK = \begin{bmatrix} -0.1 & 0 \\ -0.2 & -0.1 \end{bmatrix}. \quad (6.36)$$

Given different lower delay bound η_1, Table 6.1 lists the results of the maximum allowable upper delay bound η_3 derived from various methods including the one proposed in this paper. One can see from Table 6.1 the best results are obtained from Corollary 6.1 in this chapter than those in [115, 117–119] with less number of LMIs scalar decision variables, which show the effectiveness of the proposed method.

Example 2: Consider a spring–mass system as shown in Fig. 6.2 [120, 121], where x_1 and x_2 are two positions of masses m_1 and m_2, k_1 and k_2 are the spring constants, c denotes the viscous friction coefficient between the masses and the horizontal

Table 6.1 Calculated maximum η_3 for given η_1 (Example 1)

η_1	2	4	6	10	12	Variables
Gao et al. [117]	13	13	14	15	17	$\frac{23n^2+5n}{2}$
Zhang et al. [119]	13	13	14	17	18	$\frac{18n^2+6n}{2}$
Shao et al. [118]	17	17	18	20	21	$\frac{15n^2+5n}{2}$
Peng [115]	18	18	19	20	21	$\frac{7n^2+7n}{2}$
Corollary 6.1	19	19	20	21	22	$\frac{9n^2+7n}{2}$

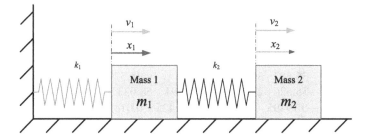

Fig. 6.2 A mass–spring system

surface. The plant noise is denoted by v_1. Assume that x_1 and x_2 are measurable and they are measured by devices with noise v_2.

Denoting $x(t) = [x_1(t)\ x_2(t)\ \dot{x}_1(t)\ \dot{x}_2(t)]^T$ and $w(t) = [v_1(t)\ v_2(t)]^T$, we obtain the following state-space model for the studied system described in Fig. 6.2:

$$\dot{x}(t) = \begin{bmatrix} 0 & 0 & 1 & 0 \\ 0 & 0 & 0 & 1 \\ -\frac{k_1+k_2}{m_1} & \frac{k_2}{m_1} & -\frac{c}{m_1} & 0 \\ \frac{k_2}{m_2} & -\frac{k_2}{m_2} & 0 & -\frac{c}{m_2} \end{bmatrix} x(t) + \begin{bmatrix} 0 & 0 \\ 0 & 0 \\ \frac{1}{m_1} & 0 \\ \frac{1}{m_2} & 0 \end{bmatrix} w(t),$$

$$y(t) = \begin{bmatrix} 1 & 0 & 0 & 0 \\ 0 & 1 & 0 & 0 \end{bmatrix} x(t) + \begin{bmatrix} 0 & 0.1 \\ 0 & 0.1 \end{bmatrix} w(t). \tag{6.37}$$

The purpose of this example is to design a networked H_∞ filter to estimate the signal $z = x_1 + x_2$, while depending on the less network bandwidth. Choose $m_1 = 1, m_2 = 0.5, k_1 = k_2 = 1, c = 0.5$ and the sampling period $h = 0.2\,\text{s}$. Discretizing the continuous-time plant model in (6.37) gives the discrete-time plant model (6.1) with the following parameters [121]

$$A = \begin{bmatrix} 0.9617 & 0.0191 & 0.1878 & 0.0012 \\ 0.0370 & 0.9629 & 0.0025 & 0.1798 \\ -0.3732 & 0.1853 & 0.8678 & 0.0179 \\ 0.3528 & -0.3553 & 0.0357 & 0.7840 \end{bmatrix},$$

$$B = \begin{bmatrix} 0.0193 & 0 \\ 0.0187 & 0 \\ 0.1890 & 0 \\ 0.1813 & 0 \end{bmatrix}, C = \begin{bmatrix} 1 & 0 & 0 & 0 \\ 0 & 1 & 0 & 0 \end{bmatrix},$$

$$D = \begin{bmatrix} 0 & 0.1 \\ 0 & 0.1 \end{bmatrix}, L = \begin{bmatrix} 1 & 1 & 0 & 0 \end{bmatrix}. \tag{6.38}$$

$$\left[\begin{array}{c|c} \bar{A}_f & \bar{B}_f \\ \hline \bar{C}_f & \bar{D}_f \end{array}\right] = \left[\begin{array}{cccc|cc} 342.99 & -250.58 & 74.769 & -120.36 & -4.0365 & 2.5009 \\ -257.96 & 189.34 & -51.758 & 87.81 & 2.427 & -1.7376 \\ -36.682 & 30.575 & 64.903 & -44.529 & 0.94781 & -0.60543 \\ -35.526 & 23.6 & -63.994 & 58.033 & -0.75845 & 0.53493 \\ \hline -0.19495 & -0.37648 & 0.0021425 & -0.014724 & 0.28171 & 0.27138 \end{array}\right]$$

$$P_3 = \begin{bmatrix} 389.55 & -287.8 & 18.507 & -85.045 \\ -287.8 & 214.35 & -9.8403 & 60.627 \\ 18.507 & -9.8403 & 71.461 & -59.169 \\ -85.045 & 60.627 & -59.169 & 63.081 \end{bmatrix}, \Phi = \begin{bmatrix} 7.8778 & -4.249 \\ -4.249 & 3.667 \end{bmatrix}$$

$$\tag{6.39}$$

Case 6.1 Give $\gamma = 3.4$, $\eta_3 = 4$ and simulation step $T = 100$. For different communication parameter δ_m, Table 6.2 lists the calculated maximum allowable $\eta_3(\delta_m)$, the number of the transmitted packets $N(\delta_m)$ through a communication network. It is known that the communication parameter δ_m should be chosen as 0.5 and Φ should be set as $\begin{bmatrix} 7.8778 & -4.2490 \\ -4.2490 & 3.6670 \end{bmatrix}$, since in this condition the number of the transmitted packets $N = 25$ is smallest under the constraint of $4 \leq \eta_3(\delta_m)$.

Case 6.2 Assume that the measurement $y(k)$ is sampled and transmitted through an IP-based network with lower delay bound $\eta_1 = 1$ and upper delay bound $\eta_3 = 4$. For given different communication parameter δ_m, applying Theorem 6.2 and Matlab LMI Toolbox, as listed in Table 6.3, one can obtain the corresponding minimum performance level $\gamma(\delta_m)$.

Table 6.2 Maximum allowable bound η_3, the number of the transmitted packets N

δ_m	0.1	0.2	0.3	0.4	0.5	0.6
$\eta_3(\delta_m)$	6	5	5	4	4	2
$N(\delta_m)$	52	45	37	31	25	21

Based on Theorem 6.2 the networked filter parameters can also be obtained. For example, for $\delta_m = 0.5$, we have (6.39). Then, from (6.39), we can finally obtain the filter with the parameters (6.40).

Assume the disturbances $v_1(k)$ and $v_2(k)$ are uniformly distributed within $[-1, 1]$, for the simulation interval $k \in [1, 20]$ (zero elsewhere). With the given simulation step $T = 100$, Table 6.3 also lists the number of the transmitted packets $N(\delta_m)$ under the proposed event-triggered communication scheme. It is clear that the number

$$
\left[\begin{array}{c|c} A_f & B_f \\ \hline C_f & D_f \end{array} \right] = \left[\begin{array}{cccc|cc} 0.59588 & -0.41812 & 1.1303 & 0.35741 & -4.0365 & 2.5009 \\ -0.22644 & 0.51261 & -0.46195 & 0.16076 & 2.427 & -1.7376 \\ -0.12994 & 0.1075 & 0.63403 & -0.38969 & 0.94781 & -0.60543 \\ 0.070356 & -0.12611 & 0.042656 & 1.176 & -0.75845 & 0.53493 \\ \hline -0.30092 & -0.45765 & 0.19264 & 0.21462 & 0.28171 & 0.27138 \end{array} \right]
$$
$$(6.40)$$

of transmitted packets generated by the proposed communication scheme is much smaller than those generated by the time-triggered scheme in [88, 121]. For instance, when $\delta_m = 0.3$ and 0.5, only 24 and 22% of the sampled data are needed to be transmitted through the communication networks, respectively. Figure 6.3 depicts the release step with the event-triggered communication scheme (*top*) and time-triggered communication scheme (*bottom*); Fig. 6.4 shows the threshold conditions of $\delta_m y^T(d_k) \Phi y(d_k) - e^T(i_k) \Phi e(i_k)$ in (6.2). Compared with all sampled data are

Table 6.3 H_∞ performance index γ and the number of the transmitted packets N

δ_m	0.1	0.3	0.5	0.7	0.9
$\gamma(\delta_m)$	2.5	3.0	3.4	3.8	4.1
$N(\delta_m)$	52	24	22	28	18

Fig. 6.3 Release steps and intervals with event-triggered and time-triggered communication schemes

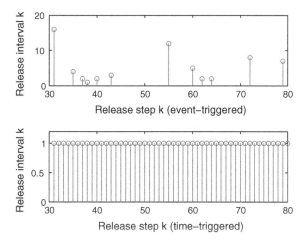

Fig. 6.4 Threshold
conditions in Fig. 6.3

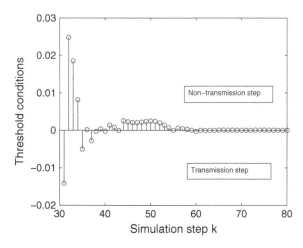

Table 6.4 Number of
transmitted packets N under
different communication
methods

Method	N
Event-triggered scheme in [114] with $\sigma = 0.1$	76
Event-triggered scheme in [114] with $\sigma = 0.3$	46
Event-triggered scheme in [114] with $\sigma = 0.5$	35
Adaptive event-triggered scheme with $\sigma_m = 0.5$	28

required to be transmitted in a time-triggered communication scheme (*bottom* in
Fig. 6.3), it is shown that only part of the sampled data are transmitted with the
proposed communication scheme (*up* in Fig. 6.3), while maintaining the desired
performance and reducing the average transmission frequency.

Case 6.3 In this case, we will show the advantage of the proposed adaptive event-
triggered scheme. For different event-triggered methods, Table 6.4 lists the numbers
of transmitted packets N though a communication network with a given simulation
step $T = 300$. It is clear that the number of transmitted packets generated by the
proposed adaptive communication scheme is much smaller than those generated by
the traditional event-triggered scheme in [68, 114, 122].

6.4 Conclusion

An H_∞ filter design method has been proposed for a discrete networked linear
system with an adaptive event-triggered communication scheme. Compared with
some existing time-triggered communication schemes, where all the sampled data are
transmitted over the communication networks regardless of the current measurement,
since the proposed adaptive event-triggering communication scheme can determine

whether or not the current measurement is necessary to be transmitted based on the error-dependent threshold between the latest transmitted measurement and the current measurement, therefore, the less limited communication bandwidth is used, while guaranteeing the desired H_∞ filter performance. Case studies have been carried out to demonstrate the effectiveness of proposed method.

Chapter 7
Codesign of Event-Triggered Communication Scheme and H_∞ Controller for NCSs

This chapter studies an event-triggered communication and H_∞ control codesign method for networked control systems (NCSs) with network-induced delays and packet dropouts. First, an event-triggered communication scheme and a sampled-state error-dependent model for NCSs are presented, in which the sensor takes sample in a periodic manner with the event-triggering condition being applied to this periodically sampled signal; and the closed-loop system with a networked state feedback controller is modeled as a time-delay system including the sampled-state error between the sampled-state at current sampling instant and the sampled-state at the latest transmitted sampling instant. Second, two communication-dependent stability criteria and a stabilization criterion are derived. Third, a codesign algorithm is developed for the triggering condition and the controller feedback gain to meet the specified performance. Apart from the fact that this is part of an overall codesign method, other unique features are (a) not like the schemes in [10, 59], where the event-triggering conditions need to be monitored continuously, here the condition is only checked at each sampling instance; and (b) it is shown that the triggering condition proposed in this chapter is in a general form, and that those conditions in [10, 59] are special cases of this form. Moreover, different from some two-step schemes [10, 59, 63], controller gains and the event-triggering condition are designed in one step in this chapter to meet H_∞ performance with respect to disturbance; and at the same time giving the maximum allowable communication delay bound (MADB) and the maximum allowable number of successive packet losses (MANSPL) [69].

This chapter is organized as follows. A general event-triggered communication scheme and NCS modeling are presented in Sect. 7.1. The main theoretical results including the stability analysis and controller design criteria are presented in Sect. 7.2. Section 7.3 provides a communication and control codesign algorithm to obtain the communication and control parameters simultaneously. Illustrative examples are given in Sect. 7.4 to demonstrate the effectiveness of proposed method. Section 7.5 concludes the chapter.

© Springer-Verlag Berlin Heidelberg 2015   97
C. Peng et al., *Communication and Control for Networked Complex Systems*,
DOI 10.1007/978-3-662-46813-5_7

7.1 Modeling of NCSs with an Event-Triggered Communication Scheme

In this section, a control system model is first proposed to link the event-triggered communication scheme with the other part of the system to be controlled; and then a completed NCS model under a unified framework is presented.

Consider a class of linear systems governed by

$$\begin{cases} \dot{x}(t) = Ax(t) + Bu(t) + B_\omega\omega(t) \\ z(t) = Cx(t) + Du(t) \end{cases} \tag{7.1}$$

where $x(t) \in \mathbb{R}^n$, $u(t) \in \mathbb{R}^m$, $\omega(t) \in L_2[0, \infty)$ and $z(t) \in \mathbb{R}^p$ are state vector, control input vector, disturbance input vector, and controlled output vector, respectively; A, B, B_ω, C, and D are constant matrices with appropriate dimensions; the initial condition of the system (7.1) is given by $x(t_0) = x_0$. Throughout this paper, it is assumed that the system (7.1) is controlled over a communication network with a networked state feedback controller, which is directly connected to the actuator through a ZOH [12]. This leads to

$$\begin{cases} \dot{x}(t) = Ax(t) + Bu(t) + B_\omega\omega(t) \\ z(t) = Cx(t) + Du(t) \\ u(t) = Kx(t_k h), \quad t \in \Omega =: [t_k h + \tau_{t_k}, t_{k+1}h + \tau_{t_{k+1}}) \end{cases} \tag{7.2}$$

where K is the state feedback controller gain to be designed; τ_{t_k} is the packet transmission delay.

Define $\eta(t) \triangleq t - i_k h$, $t \in \Omega_\ell$, where Ω_ℓ is sampling-interval-like subsets $\Omega_\ell = [i_k h + \tau_{t_k}, i_k h + h + \tau_{t_{k+1}})$ [16, 69], i.e., $\Omega = \cup\Omega_\ell$, where $i_k h = t_k h + \ell h$, $\ell = 0, \ldots, t_{k+1} - t_k - 1$. It is clear that $\eta(t)$ is a piecewise-linear function satisfying

$$0 < \eta_1 \le \eta(t) \le h + \bar{\tau} = \eta_3, \quad t \in \Omega_\ell \tag{7.3}$$

where $\eta_1 = inf_\ell\{\tau_{t_k}\}$, $\eta_3 = h + sup_\ell\{\tau_{t_{k+1}}\} = h + \bar{\tau}$; h and $\bar{\tau}$ are the sampling period and the maximum allowable upper communication delay bound, respectively. By making use of the event-triggered communication scheme and model methods provided in Chap. 2, control of (7.2) can be written as

$$u(t) = K(x(t - \eta(t)) - e(i_k h)), \quad t \in \Omega_\ell \tag{7.4}$$

Combining (7.2) and (7.4) leads to a sampled-state error-dependent closed-loop NCS model:

$$\begin{cases} \dot{x}(t) = Ax(t) + BK(x(t - \eta(t)) - e(i_k h)) + B_\omega\omega(t) \\ z(t) = Cx(t) + DK(x(t - \eta(t)) - e(i_k h)), \quad t \in \Omega_\ell \end{cases} \tag{7.5}$$

where the initial condition of the state $x(t)$ is: $x(t) = \phi(t), t \in [t_0 - \eta_3, t_0], \phi(t_0) = x_0$, and $\phi(t)$ is a continuous function on $[t_0 - \eta_3, t_0]$.

The purpose of this chapter is to provide an event-triggered communication and H_∞ control codesign method such that the system (7.5) is asymptotically stable with an H_∞ disturbance attenuation level γ, i.e.,

 (i) System (7.5) with $\omega(t) = 0$ is asymptotically stable; and
 (ii) Under the zero initial condition, $\|z(t)\|_2 < \gamma \|\omega(t)\|_2$, for any nonzero $\omega(t) \in$ Ł$_2[0, \infty)$ and a prescribed $\gamma > 0$.

7.2 H_∞ Stability Analysis and Controller Design

In this section, two stability theorems are first developed for the system (7.5) with network-induced delay and with or without data dropout, respectively. Then, Theorem 7.3 is presented in Sect. 7.2.3 which lays the foundation for the codesign algorithm presented in Sect. 7.3.

7.2.1 H_∞ Stability Analysis with Network-Induced Delay

Theorem 7.1 *For some given positive constants η_1, η_3, γ and δ_1, a matrix K, under the event-triggered communication scheme* (2.13), *the system* (7.5) *is asymptotically stable with an H_∞ performance index γ for the disturbance attention, if there exist real matrices $P > 0$, $\Phi > 0$, $S > 0$, $R_j > 0$ ($j = 1, 2, 3$), $\begin{bmatrix} Q_1 & * \\ Q_3 & Q_2 \end{bmatrix} > 0$, $\begin{bmatrix} R_i & * \\ U_i & R_i \end{bmatrix} > 0$ ($i = 2, 3$), matrices Q_3, U_2, U_3 with appropriate dimensions, such that*

$$\begin{bmatrix} \Pi^i_{11} & * \\ \Pi_{21} & \Pi_{22} \end{bmatrix} < 0, i = 2, 3 \tag{7.6}$$

where

$$\Pi^i_{11} = \begin{bmatrix} F_{11} & * & * & * & * & * & * \\ R_1 & F_{22} & * & * & * & * & * \\ K^T B^T P & F^i_{32} & F^i_{33} & * & * & * & * \\ 0 & F^i_{42} & F^i_{43} & F_{44} & * & * & * \\ 0 & 0 & F^i_{53} & F^i_{54} & F_{55} & * & * \\ -K^T B^T P & 0 & \delta_1 \Phi & 0 & 0 & F_{66} & * \\ B^T_\omega P & 0 & 0 & 0 & 0 & 0 & -\gamma^2 I \end{bmatrix}$$

$$\Pi_{21} = col\{\eta_1 R_1 \mathscr{I}_1, (\eta_2 - \eta_1) R_2 \mathscr{I}_1, (\eta_3 - \eta_2) R_3 \mathscr{I}_1, \mathscr{I}_2\}$$

$$\Pi_{22} = diag\{-R_1, -R_2, -R_3, -I\}$$

with

$$F_{11} = PA + A^T P + S - R_1 \qquad F_{32}^2 = R_2 - U_2$$
$$F_{22} = Q_1 - S - R_1 - R_2 \qquad F_{42}^2 = Q_3 + U_2$$
$$F_{33}^2 = \delta_1 \Phi - 2R_2 + U_2^T + U_2 \quad F_{43}^2 = R_2 - U_2$$
$$F_{44} = Q_2 - Q_1 - R_2 - R_3 \qquad F_{53}^2 = 0$$
$$F_{55} = -Q_2 - R_3 \qquad\qquad F_{54}^2 = R_3 - Q_3$$
$$F_{33}^3 = \delta_1 \Phi - 2R_3 + U_3^T + U_3 \quad F_{42}^3 = Q_3 + R_2$$
$$F_{43}^3 = R_3 - U_3^T \qquad\qquad F_{32}^3 = 0$$
$$F_{53}^3 = R_3 - U_3 \qquad\qquad F_{54}^3 = U_3 - Q_3$$
$$\mathscr{I}_1 = [A, 0, BK, 0, 0, -BK, B_\omega] \; F_{66} = \delta_1 \Phi - \Phi$$
$$\mathscr{I}_2 = [C, 0, DK, 0, 0, -DK, 0]$$

Proof Construct a Lyapunov-Krasovskii functional candidate as

$$V(t, x_t) = x^T(t)Px(t) + \int_{t-\eta_1}^t x^T(v)Sx(v)dv$$

$$+ \sum_{i=1}^3 (\eta_i - \eta_{i-1}) \int_{-\eta_i}^{-\eta_{i-1}} \int_{t+s}^t \dot{x}^T(v)R_i\dot{x}(v)dvds$$

$$+ \int_{t-\rho}^t \xi^T(v)Q\xi(v)dv, \, t \in \Omega \tag{7.7}$$

where $P > 0, S > 0, R_i > 0 \; (i = 1, 2, 3), \eta_0 = 0, \rho \triangleq \frac{\eta_3 - \eta_1}{2}, \eta_2 \triangleq \frac{\eta_1 + \eta_3}{2}$ and

$$Q \triangleq \begin{bmatrix} Q_1 & * \\ Q_3 & Q_2 \end{bmatrix} > 0, \quad \xi(v) \triangleq \begin{bmatrix} x(v - \eta_1) \\ x(v - \eta_2) \end{bmatrix}$$

Taking the time derivative of $V(t, x_t)$ with respect to t along the trajectory of (7.5) yields

$$\dot{V}(t, x_t) = 2x^T(t)P\dot{x}(t) - x^T(t - \eta_1)Sx(t - \eta_1)$$
$$+ \xi^T(t)Q\xi(t) - \xi^T(t - \rho)Q\xi(t - \rho)$$
$$+ x^T(t)Sx(t) + \sum_{i=1}^3 \dot{x}^T(t)(\eta_i - \eta_{i-1})^2 R_i\dot{x}(t)$$
$$- \sum_{i=1}^3 (\eta_i - \eta_{i-1}) \int_{t-\eta_i}^{t-\eta_{i-1}} \dot{x}^T(v)R_i\dot{x}(v)dv \tag{7.8}$$

Define $\Delta_j \triangleq \begin{bmatrix} -R_j & * \\ R_j & -R_j \end{bmatrix}, j = 1, 2, 3.$ For $\eta(t) \in [\eta_{i-1}, \eta_i), i = 2, 3,$ applying Lemmas 2.2 and 2.3 in Chap. 2 yields

$$-\sum_{j=1,j\neq i}^{3} (\eta_j - \eta_{j-1}) \int_{t-\eta_j}^{t-\eta_{j-1}} \dot{x}^T(v) R_j \dot{x}(v) dv$$

$$\leq \sum_{j=1,j\neq i}^{3} \begin{bmatrix} x(t-\eta_{j-1}) \\ x(t-\eta_j) \end{bmatrix}^T \Delta_j \begin{bmatrix} x(t-\eta_{j-1}) \\ x(t-\eta_j) \end{bmatrix} \tag{7.9}$$

$$-(\eta_i - \eta_{i-1}) \int_{t-\eta_i}^{t-\eta_{i-1}} \dot{x}^T(v) R_i \dot{x}(v) dv \leq -\sum_{j=1}^{4} \mathfrak{S}_j^i \tag{7.10}$$

$$\mathfrak{S}_1^i = [\rho_{1,i}^T(t) - \rho_2^T(t)] R_i [\rho_{1,i}(t) - \rho_2(t)]$$
$$\mathfrak{S}_2^i = [\rho_2^T(t) - \rho_{3,i}^T(t)] R_i [\rho_2(t) - \rho_{3,i}(t)]$$
$$\mathfrak{S}_3^i = [\rho_{1,i}^T(t) - \rho_2^T(t)] U_i^T [\rho_2(t) - \rho_{3,i}(t)]$$
$$\mathfrak{S}_4^i = [\rho_2^T(t) - \rho_{3,i}^T(t)] U_i [\rho_{1,i}(t) - \rho_2(t)]$$

From the communication scheme (2.1), for $i_k h \in (t_k h, t_{k+1} h)$, it is clear that

$$e^T(i_k h) \Phi e(i_k h) \leq \delta_1 x^T(t_k h) \Phi x(t_k h) \tag{7.11}$$

Utilizing (7.9) and (7.10) to deal with the integral items in (7.8), one can get

$$\dot{V}(t, x_t) = \dot{V}(t, x_t) + e^T(i_k h) \Phi e(i_k h) - e^T(i_k h) \Phi e(i_k h)$$
$$\leq \rho^T(t)(\Pi_{11}^i - \Pi_{21}^T \Pi_{22}^{-1} \Pi_{21}) \rho(t)$$
$$-z^T(t) z(t) + \gamma^2 \omega^T(t) \omega(t) \tag{7.12}$$

where $\Pi_{11}^i, i = 2, 3, \Pi_{21}$ and Π_{22} being defined in Theorem 7.1 and

$$\rho^T(t) = [x^T(t), x^T(t-\eta_1), x^T(t-\eta(t)),$$
$$x^T(t-\eta_2), x^T(t-\eta_3), e_k^T(i_k h), \omega^T(t)].$$

First, we show that the system (7.5) is asymptotically stable. Using the Lyapunov-Krasovskii functional (7.7), we have

$$\dot{V}(t, x_t) \leq \zeta^T(t)(\hat{\Pi}_{11}^i - \hat{\Pi}_{21}^T \hat{\Pi}_{22}^{-1} \hat{\Pi}_{21}) \zeta(t) \tag{7.13}$$

where

$$\zeta^T(t) = [x^T(t), x^T(t - \eta_1), x^T(t - \eta(t)), x^T(t - \eta_2), x^T(t - \eta_3), e^T(i_k h)]$$

and

$$\hat{\Pi}_{11}^i = \begin{bmatrix} F_{11} & * & * & * & * & * \\ R_1 & F_{22} & * & * & * & * \\ K^T B^T P & F_{32}^i & F_{33}^i & * & * & * \\ 0 & F_{42}^i & F_{43}^i & F_{44} & * & * \\ 0 & 0 & F_{53}^i & F_{54}^i & F_{55} & * \\ -K^T B^T P & 0 & \delta_1 \Phi & 0 & 0 & F_{66} \end{bmatrix}$$

$$\hat{\Pi}_{21} = col\{\eta_1 R_1 \hat{\mathscr{I}}_1, (\eta_2 - \eta_1)R_2 \hat{\mathscr{I}}_1, (\eta_3 - \eta_2)R_3 \hat{\mathscr{I}}_1\}$$
$$\hat{\Pi}_{22} = -diag\{R_1, R_2, R_3\}, \quad \hat{\mathscr{I}}_1 = [A, 0, BK, 0, 0, -BK]$$

Using Schur complements (Lemma 2.1 in Chap. 2), from (7.6), we have $\hat{\Pi}_{11}^i + \hat{\Pi}_{21}^T \hat{\Pi}_{22}^{-1} \hat{\Pi}_{21} < 0$, from which $\dot{V}(t, x_t) < -\varsigma \|x(t)\|^2 < 0$ for a sufficiently small $\varsigma > 0$. This prove that the system (7.5) is asymptotically stable with $\omega(t) = 0$.

Next, under the zero initial condition, we show that $\|z(t)\|_2 < \gamma \|w(t)\|_2$ for all nonzero $\omega(t) \in \text{Ł}_2[0, \infty)$. Since the condition of (7.6) implies that

$$\Pi_{11}^i - \Pi_{21}^T \Pi_{22}^{-1} \Pi_{21} < 0, \quad i = 2, 3 \tag{7.14}$$

Then from (7.12) and (7.14), we have

$$\dot{V}(t, x_t) < -z^T(t)z(t) + \gamma^2 \omega^T(t)\omega(t) \tag{7.15}$$

Under zero initial condition, integrating both sides of (7.15) from t_0 to t and letting $t \to \infty$, yields

$$\int_{t_0}^{\infty} z(s)^T z(s) ds < \int_{t_0}^{\infty} \gamma^2 w(s)^T w(s) ds \tag{7.16}$$

which implies that $\|z(t)\|_2 < \gamma \|w(t)\|_2$. This completes the proof.   □

7.2.2 H_∞ Stability Analysis with Network-Induced Delays and Packet Loss

In this section, a new theorem is developed for achieving the H_∞ performance while there is network-induced delays and packet loss in the signal transfer. Based on the communication scheme (2.16) proposed in Chap. 2, the following result is derived.

Theorem 7.2 *For some given positive constants η_1, η_3, γ, h, δ_i ($i = 1, 2, 3$), a matrix K, under the event-triggered communication scheme (2.16), the system (7.5) is asymptotically stable with an H_∞ performance index γ for the disturbance attention, if there exist real matrices $P > 0$, $\Phi > 0$, $S > 0$, $R_j > 0$ ($j = 1, 2, 3$), $\begin{bmatrix} Q_1 & * \\ Q_3 & Q_2 \end{bmatrix} > 0$, $\begin{bmatrix} R_i & * \\ U_i & R_i \end{bmatrix} > 0$ ($i = 2, 3$), matrices Q_3, U_2, U_3 with appropriate dimensions, such that (7.6) holds, and the number of successive packet losses d_k satisfies*

$$d_k \le d_{MANSPL} \triangleq \left\lfloor \log_{(1+\sqrt{\delta_2})(1+\varepsilon)} \frac{1 + \sqrt{\delta_1}}{1 + \sqrt{\delta_2}} \right\rfloor \tag{7.17}$$

where $\varepsilon = |Ah| + \left| \frac{hBK}{1-\sqrt{\delta_1}} \right| + |\delta_3 hB_\omega|$ and $\lfloor \diamond \rfloor$ gives the largest integer smaller than or equal to $\diamond$.

Proof Consider a successful broadcast release interval $[t_k h, t_{k+1} h)$, and assume that in this interval, there exists the number of d_k unsuccessfully transmitted broadcast packets, that is

$$t_k = b_0 < b_1 < b_2 < \cdots < b_{d_k} < b_{d_k+1} = t_{k+1} \tag{7.18}$$

For $l = 0, 1, \ldots, d_k$, applying the communication scheme (2.16) yields

$$|x(b_{l+1}h - h) - x(b_l h)| \le \sqrt{\delta_2} |x(b_l h)| \tag{7.19}$$

It is clear that

$$|x(b_{l+1}h - h)| \le |x(b_{l+1}h - h) - x(b_l h)| + |x(b_l h)|$$
$$\le (1 + \sqrt{\delta_2}) |x(b_l h)| \tag{7.20}$$

For the closed-loop system (7.5) with a disturbance input vector $\omega(t) \in \mathcal{L}_2[0, \infty)$, one can derived that the step increment of the system' state is bounded, i.e.,

$$|x(b_{l+1}h) - x(b_{l+1}h - h)| \le \varepsilon |x(b_{l+1}h - h)| \tag{7.21}$$

where $\varepsilon > 0$ is related to the system's modeling parameters and sampling period h. From (7.20) and (7.21), we have

$$|x(b_{l+1}h) - x(b_{l+1}h - h)| \le \varepsilon(1 + \sqrt{\delta_2}) |x(b_l h)| \tag{7.22}$$

Considering (7.19) and (7.22) together, for $t \in [b_{d_k}h, b_{d_k+1}h)$, the state error between two successfully transmitted sampling can be evolved as

$$|x(t) - x(t_k h)| \leq |x(t) - x(b_{d_k}h)| + \left| \sum_{l=0}^{d_k-1} (x(b_{l+1}h - h) - x(b_l h)) \right|$$

$$+ \left| \sum_{l=0}^{d_k-1} (x(b_{l+1}h) - x(b_{l+1}h - h)) \right|$$

$$\leq \sum_{l=0}^{d_k} \sqrt{\delta_2} |x(b_l h)| + \sum_{l=0}^{d_k-1} |x(b_{l+1}h) - x(b_{l+1}h - h)|$$

$$\leq \sum_{l=0}^{d_k} \sqrt{\delta_2} |x(b_l h)| + \sum_{l=0}^{d_k-1} \varepsilon(1 + \sqrt{\delta_2}) |x(b_l h)| \qquad (7.23)$$

From (7.19) and (7.22), it is clear that

$$|x(b_{l+1}h)| \leq |x(b_{l+1}h) - x(b_{l+1}h - h)|$$
$$+ |x(b_{l+1}h - h) - x(b_l h)| + |x(b_l h)|$$
$$\leq (1 + \sqrt{\delta_2}) |x(b_l h)| + \varepsilon(1 + \sqrt{\delta_2}) |x(b_l h)|$$
$$\leq [(1 + \varepsilon)(1 + \sqrt{\delta_2})]^{l+1} |x(t_k h)| \qquad (7.24)$$

Using (7.24) to deal with the term of $|x(b_l h)|$ in (7.23), yields

$$|x(t) - x(t_k h)| \leq \sum_{l=0}^{d_k} \sqrt{\delta_2} |x(b_l h)| + \sum_{l=0}^{d_k-1} \varepsilon(1 + \sqrt{\delta_2}) |x(b_l h)|$$

$$\leq \sum_{l=0}^{d_k} \sqrt{\delta_2}[(1 + \varepsilon)(1 + \sqrt{\delta_2})]^l |x(t_k h)|$$

$$+ \sum_{l=0}^{d_k-1} \varepsilon(1 + \sqrt{\delta_2})[(1 + \varepsilon)(1 + \sqrt{\delta_2})]^l |x(t_k h)|$$

$$= ((1 + \sqrt{\delta_2})^{d_k+1}(1 + \varepsilon)^{d_k} - 1) |x(t_k h)| \qquad (7.25)$$

Considering (7.17) and (7.25) together, we have

$$|x(t) - x(t_k h)| \leq \sqrt{\delta_1} |x(t_k h)| \qquad (7.26)$$

From (7.26), one can see that the event-triggered communication condition (2.13) in Theorem 7.1 is ensured by the communication scheme (2.16), this reveals that

Theorem 7.2 can be readily derived from Theorem 7.1 if the communication scheme (2.16) is utilized. This completes the proof.   □

The maximum allowable number of successive packet losses d_{MANSPL} in (7.17) is a nonnegative integer. This implies that $\delta_2 \leq \delta_1$. As a special case, when $\delta_2 = \delta_1$, $d_{MANSPL} = 0$, i.e., packet loss is not permitted. $\delta_2 \leq \delta_1$ also means to lower the threshold of the triggering condition which causes more packets being transmitted. This is necessary since δ_1 in (2.13) assumes no packet loss.

7.2.3 H_∞ Controller Design

The next theorem lays the foundation for the codesign algorithm presented in the next section. The proof is based on Theorem 7.2, and it is omitted here.

Theorem 7.3 *For some given positive constants η_1, η_3, h, γ, δ_i ($i = 1, 2, 3$), under the given communication scheme (2.16), the system (7.5) is asymptotically stable with an H_∞ performance index γ for disturbance attention and a state feedback gain $K = YX^{-1}$, if the number of successive packet losses d_k satisfies (7.17) and there exist real matrices $X > 0$, $\tilde{S} > 0$, $\tilde{\Phi} > 0$, $\tilde{R}_j > 0$ ($j = 1, 2, 3$), $\begin{bmatrix} \tilde{Q}_1 & * \\ \tilde{Q}_3 & \tilde{Q}_2 \end{bmatrix} > 0$,
$\begin{bmatrix} \tilde{R}_i & * \\ \tilde{U}_i & \tilde{R}_i \end{bmatrix} > 0$ ($i = 2, 3$), matrices $\tilde{Q}_3$, $\tilde{U}_2$, $\tilde{U}_3$, Y with appropriate dimensions, such that*

$$\begin{bmatrix} \tilde{\Pi}^i_{11} & * \\ \tilde{\Pi}_{21} & \tilde{\Pi}_{22} \end{bmatrix} < 0, \quad i = 2, 3 \tag{7.27}$$

where

$$\tilde{\Pi}^i_{11} = \begin{bmatrix} \tilde{F}_{11} & * & * & * & * & * & * \\ \tilde{R}_1 & \tilde{F}_{22} & * & * & * & * & * \\ Y^T B^T & \tilde{F}^i_{32} & \tilde{F}^i_{33} & * & * & * & * \\ 0 & \tilde{F}^i_{42} & \tilde{F}^i_{43} & \tilde{F}_{44} & * & * & * \\ 0 & 0 & \tilde{F}^i_{53} & \tilde{F}^i_{54} & \tilde{F}_{55} & * & * \\ -Y^T B^T & 0 & 0 & 0 & 0 & \tilde{F}_{66} & * \\ B^T_\omega & 0 & 0 & 0 & 0 & 0 & -\gamma^2 I \end{bmatrix}$$

$$\tilde{\Pi}_{21} = col\{\eta_1 \tilde{\mathscr{I}}_1, (\eta_2 - \eta_1)\tilde{\mathscr{I}}_1, (\eta_3 - \eta_2)\tilde{\mathscr{I}}_1, \tilde{\mathscr{I}}_2\}$$
$$\tilde{\Pi}_{22} = -diag\{X\tilde{R}_1^{-1}X, X\tilde{R}_2^{-1}X, X\tilde{R}_3^{-1}X, I\}$$

with

$$\tilde{F}_{11} = AX + XA^T + \tilde{S} - \tilde{R}_1 \qquad \tilde{F}^2_{32} = \tilde{R}_2 - \tilde{U}_2$$
$$\tilde{F}_{22} = \tilde{Q}_1 - \tilde{S} - \tilde{R}_1 - \tilde{R}_2 \qquad \tilde{F}^2_{42} = \tilde{Q}_3 + \tilde{U}_2$$
$$\tilde{F}^2_{33} = \delta_1 \tilde{\Phi} + \tilde{U}^T_2 + \tilde{U}_2 - 2R_2 \qquad \tilde{F}^2_{43} = \tilde{R}_2 - \tilde{U}_2$$
$$\tilde{F}_{44} = \tilde{Q}_2 - \tilde{Q}_1 - \tilde{R}_2 - \tilde{R}_3 \qquad \tilde{F}^2_{53} = 0$$
$$\tilde{F}_{55} = -\tilde{Q}_2 - \tilde{R}_3 \qquad \tilde{F}^2_{54} = \tilde{R}_3 - \tilde{Q}_3$$
$$\tilde{F}^3_{33} = \delta_1 \tilde{\Phi} + \tilde{U}^T_3 + \tilde{U}_3 - 2\tilde{R}_3 \qquad \tilde{F}^3_{42} = \tilde{Q}_3 + \tilde{R}_2$$
$$\tilde{F}^3_{43} = \tilde{R}_3 - \tilde{U}^T_3 \qquad \tilde{F}^3_{32} = 0$$
$$\tilde{F}^3_{53} = \tilde{R}_3 - \tilde{U}_3 \qquad \tilde{F}^3_{54} = \tilde{U}_3 - \tilde{Q}_3$$
$$\tilde{\mathscr{I}}_1 = [AX, 0, BY, 0, 0, -BY, B_\omega] \, \tilde{F}_{66} = \delta_1 \tilde{\Phi} - \tilde{\Phi}$$
$$\tilde{\mathscr{I}}_2 = [CX, 0, DY, 0, 0, -DY, 0]$$

Proof Define $X = P^{-1}, X\Phi X = \tilde{\Phi}, XSX = \tilde{S}, XR_j X = \tilde{R}_j, XQ_j X = Q_j, j = 1, 2, 3,$ $XU_i X = \tilde{U}_i, i = 2, 3,$ and $Y = KX$. Then, pre and postmultiply both sides of leftmost matrix of (7.6) with diag$(X, X, X, X, X, X, I, I, I, I)$ and its transpose, respectively. Based on the above symbol definitions, (7.27) can be really obtained from Theorem 7.2. This completes the proof. $\qquad\qquad\qquad\qquad\qquad\qquad\qquad\qquad\qquad\qquad\qquad\qquad\square$

7.3 Codesign Algorithm

The parameters δ_1, δ_2, Φ in the event-triggering scheme and the controller gain K are coupled together in Theorem 7.3; and at the same time, the control performance and the network resource usage are related to these parameters. Thus, it is necessary to develop an algorithm to obtain these parameters simultaneously for the desired H_∞ performance while using less network resource.

For convenience, the average transmission time $\tilde{T}$ is defined as the ratio of a given period of time T to the number of transmitted sampled data.

Algorithm 1: Find the communication parameters δ_1, δ_2, Φ, and the controller gain K

Step 1.   For the given d_{MANSPL} and $\bar{\tau}$, set $\delta_1 = \delta_1 + \lambda$, where λ is the step increment of δ_1, δ_1 is initially set to zero, $\delta_1 \in [0, 1)$.

Step 2.   For a given δ_1 in Step 1, if there exists a feasible solution satisfying LMIs defined in the CCL algorithm, go to Step 3; Otherwise go to Step 1.

Step 3.   Use the Matlab LMI Toolbox and the CCL algorithm to find the maximum $\eta_3(\delta_1)$, and the corresponding $\Phi(\delta_1), K(\delta_1)$ based on Theorem 8.4.

Step 4.   Use the given d_{MANSPL} and the current δ_1 to calculate the maximum $\delta_2(\delta_1, d_{MANSPL})$ under the constraint of (7.17). If $\delta_2(\delta_1, d_{MANSPL}) \leq 0$, go to Step 1.

Step 5. Set the sampling period $h = \eta_3(\delta_1) - \bar{\tau}$. If $h \leq 0$, go to Step 1; Else based on $\Phi(\delta_1)$ and $K(\delta_1)$ in Step 3, and $\delta_2(\delta_1, d_{MANSPL})$ in Step 4, set a simulation time T, and based on the proposed communication scheme, use Matlab/Simulink to find the average transmission time $\tilde{T}$.

Step 6. Go to Step 1 for another value of δ_1, if feasible, find another $\tilde{T}$ for this particular δ_1, until $\delta_1 \geq 1$ when the search is terminated.

Notice that, in the mathematical theorems, for a given set of parameters and the performance requirement, one can find d_{MANSPL} and $\bar{\tau}$; whereas in the design process, one decide d_{MANSPL} and $\bar{\tau}$ first based on the knowledge of the network being used, then find the parameters using the above algorithm. In addition, since $\delta_1 = \delta_1 + \lambda$ and $\delta_1 \in [0, 1)$ in Step 1, Algorithm 1 terminates in a finite number of steps $M \in \mathbb{N}^+$ and $M \leq 1/\lambda$.

7.4 Numerical Examples

This section employs two examples to demonstrate the effectiveness of the proposed approach. They show that the system performance can be maintained while using less network bandwidth and allowing a degree of data dropouts in the communication.

Example 1: Consider the following system controlled over a network

$$\dot{x}(t) = \begin{bmatrix} 0 & 1 \\ 0 & -0.1 \end{bmatrix} x(t) + \begin{bmatrix} 0 \\ 0.1 \end{bmatrix} u(t) \tag{7.28}$$

The non-networked controller is designed as $u(t) = -[\, 3.75 \;\; 11.5\,] x(t)$.

With the given simulation time $T = 30\,\text{s}$, and $\delta_1 = 0.01, 0.09, 0.25$, and 0.49, respectively, Table 7.1 shows the obtained maximum allowable η_3 based on Theorem 7.1. Moreover, based on Theorem 7.2 ($\delta_2 = 0.01, \varepsilon = 0.01$), the average transmission time $\tilde{T}$ and the maximal allowable number of successive dropouts d_{MANSD} are also listed in Table 7.1. Compared with 100 % sampled data should be transmitted by the time-triggered scheme in [12, 15, 98, 99], one can see that not all sampled data are transmitted by the proposed communication scheme (2.13), for example, with $\delta_1 = 0.49$, and $\delta_2 = 0.01$, not more than 51 % sampled data are transmitted, which shows the effectiveness of proposed communication scheme.

Table 7.1 MANSD and average transmission period $\tilde{T}$ with different δ_1

δ_1	0.01	0.09	0.25	0.49
η_3 (Theorem 7.1)	1.06	0.97	0.79	0.52
$\tilde{T}$ (Theorem 7.2)	1.25	1.48	1.43	1.01
d_{MANSD}	0	1	2	4

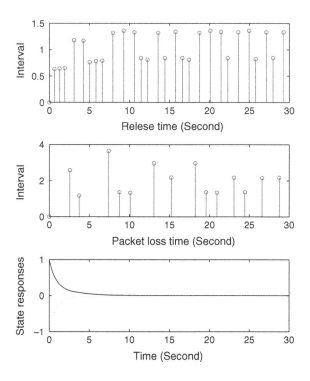

Fig. 7.1 Communication condition, data dropout, and state responses with an event-triggered scheme (Example 1)

With given $\delta_1 = 0.49, \varepsilon = 0.01, \delta_2 = 0.01$ and $x(0) = [1, -1]^T$ the release times, data dropouts, and system states are shown in Fig. 7.1. As expected, the release times are triggered when the condition defined in (2.13) is violated. Moreover, although some packets are dropped in the communication, the system is still asymptotically state with the proposed communication scheme.

Furthermore, it is clear that if set $\delta_1 = 0$ in (2.13), the event-triggered scheme proposed in this chapter includes the time-triggered communication scheme employed in [12, 98, 99, 123] as a special case. Using the stability criteria in the open literature, and Theorem 7.1 in this chapter ($\delta_1 = 0$), Table 7.2 lists the maximum allowable delay bounds (MADB), which guarantees the stability of the system (7.28) controlled over a network. It is seen from Table 7.2 that the MADB (1.07) obtained in this chapter outperforms those in [12, 98, 99, 123].

Example 2: Consider the example taken from [59, 63]. The plant (7.2) is governed by

$$\dot{x}(t) = \begin{bmatrix} 0 & 1 \\ -2 & 3 \end{bmatrix} x(t) + \begin{bmatrix} 0 \\ 1 \end{bmatrix} u(t) \tag{7.29}$$

Table 7.2 Comparisons of MADB

Criterion	[98]	[12]	[99]	[123]	Theorem 7.1
MATI	0.78	0.86	0.94	1.04	1.07

Table 7.3 Lower bound of the inter-event times and allowable upper bound of δ_1 (Example 2)

δ_1	0.003	0.0273	0.0588	0.10
[124]	0.0318	Fail	Fail	Fail
[63]	–	0.0840	0.1136	Fail
Theorem 7.1	0.2141	0.1922	0.1602	0.1125

For a fair comparison, we set $\Phi = I$ in the communication scheme (2.13) and $\varepsilon = 0$ in the communication scheme in [63], and choose $K = [1, -4]$ as those in [59, 63]. From (2.13) and (2.16), it is observed that the inter-event time in this study is the sampling period h. Under the zero network-induced delays assumption, Table 7.3 lists the results of the minimum inter-event times and the maximum allowable δ_1 derived from the study of Mazo et al. [124], Donkers and Heemels [63], and Theorem 7.1 of this chapter, respectively. It is clearly seen from Table 7.3 that Theorem 7.1 gives the largest inter-event times. Moreover, to guarantee the desired performance, the conditions of $\delta_1 \leq 0.003$ and $\delta_1 \leq 0.0588$ should be satisfied in [124] and [63], respectively. However, when $\delta_1 = 0.10$, the communication scheme employed in the study still works well. We can therefore conclude that taking use of the approach in this chapter increases the allowable minimum inter-event times and allows a larger communication parameter δ_1 with respect to the approaches in [59, 63, 124].

7.5 Conclusion

A combined event-triggering condition and controller feedback gain codesign method for NCSs is presented in this chapter. This design method with the proposed algorithm maintains the desired system H_∞ performance, takes into account network-induced delays and packet loss in the networked signal transfer, and makes a better use of network resources. The theoretical background of the proposed design method is a novel Lyapunov-Krasovskii functional and the three theorems proved in this chapter. The application of the design method and some of its advantages over other existing methods are demonstrated by two numerical examples.

Chapter 8
Self-triggered Sampling Scheme for NCSs

In this chapter, a self-triggered sampling scheme for an NCS is proposed by considering network-induced delays and data dropouts simultaneously. The next sampling instant will be estimated based on the desired performance; the latest transmitted time-stamped control packets; and the network-induced delays and data dropouts. In the proposed self-triggered scheme, the state of the system does not need to be sampled except that a predicted evolution of a function of the state is larger than a certain threshold. Thus, the transmission is aperiodic and reduces the communication burden while maintaining the desired H_∞ performance. Compared with some existing continuous event-triggered sampling schemes [10, 59], the proposed scheme will dynamically adjust the sampling interval without depending on the continuous measurement and online estimation of an event-triggered condition.

This chapter is organized as follows. Section 8.1 presents the system model and problem description. Section 8.2 proposes an error-dependent condition estimation method and a self-triggered sampling scheme under consideration of network-induced delays and data dropouts simultaneously. Stability analysis is given in Sect. 8.3. An example is given in Sect. 8.4 to demonstrate the effectiveness of proposed sampling scheme. Section 8.5 concludes the chapter.

8.1 System and Problem Description

Consider the system described by

$$\begin{cases} \dot{x}(t) = Ax(t) + Bu(t) + B_1 w(t) \\ z(t) = Cx(t) + Du(t) \end{cases} \tag{8.1}$$

where $x(t) \in \mathbb{R}^n$, $u(t) \in \mathbb{R}^m$, and $z(t) \in \mathbb{R}^p$ are the state vector, the control input vector, and the controlled output vector, respectively; A, B, B_1, C, and D are constant matrices of appropriate dimensions; $w(t) \in \mathbb{R}^q$ is an exogenous disturbance in an L_2 space. The initial condition of the system (8.1) is given by $x(t_0) = x_0$. In this

© Springer-Verlag Berlin Heidelberg 2015
C. Peng et al., *Communication and Control for Networked Complex Systems*,
DOI 10.1007/978-3-662-46813-5_8

chapter, it is assumed that the sensor and the controller are connected by a wireless network [79].

For ease of exposition, we need to introduce the following notations:

- b_k: The instant when the sensor gets the sampled-data from the plant determined by a self-triggered sampling scheme.
- t_k: The successfully transmitted sampling instant with a time-stamp.
- p_k: The predicted instant by a predictive agent while ensuring the desired performance.
- τ_{t_k}: The network-induced delay, which is the time from the instant t_k to the instant when the actuator accepts the time-stamped control packet.
- $f_k \triangleq t_k + \tau_{t_k}$: The instant when the time-stamped control packet arrives at the actuator.

Let $T_k = b_{k+1} - b_k$. One can see that T_k can be interpreted as a time-varying sampling period. If $\{t_k\}_{k=0}^{\infty} = \{b_k\}_{k=0}^{\infty}$, there is no data dropout in the transmission; if $\{t_k\}_{k=0}^{\infty} \subset \{b_k\}_{k=0}^{\infty}$, there are data dropouts in the transmission.

In this chapter, we assume that network-induced delay τ_{t_k} is time-varying and has the lower bound τ_m and upper bound τ_M, which means that $0 < \tau_m \leq \tau_{t_k} \leq \tau_M$.

The conceptual diagram of an NCS with a self-triggered sampling scheme is illustrated in Fig. 8.1, where the dashed line means that the event of the predictive agent to estimate the next sampling is triggered by the event of the control packet arrived at the actuator, and

- the sensor is self-triggered; the controller and the actuator are event-triggered; the sampled-data of the plant is transmitted with a single packet;
- the role of the self-triggered sensor is to estimate the next sampling instant b_{k+1}, which is determined by the computed p_{k+1} by the predictive agent, the allowable number of data dropouts and the network-induced delay.
- the zero-order holder (ZOH) at the actuator is used to select and store the latest control packet.

Fig. 8.1 A diagram of an NCS with a self-triggered sampling scheme

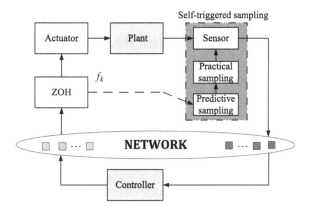

In this chapter, we consider the case of state feedback control. Then $u(t)$ is a piecewise function

$$u(t^+) = Kx(t_k), \quad t \in [f_k, f_{k+1}) \tag{8.2}$$

where $u(t^+) = \lim_{\hat{t} \to t+0} u(\hat{t})$, $t \in [f_k, f_{k+1})$ is the holding time of the ZOH, and K is the given state feedback controller gain.

From (8.1) and (8.2), we have the following closed-loop system

$$\begin{cases} \dot{x}(t) = Ax(t) + BKx(t_k) + B_1 w(t) \\ z(t) = Cx(t) + DKx(t_k), \quad t \in [f_k, f_{k+1}) \end{cases} \tag{8.3}$$

The purpose of this chapter is to design a self-triggered sampling scheme by taking network-induced delays and data dropouts into consideration such that the system (8.3) is asymptotically stable with an H_∞ disturbance attenuation level γ, i.e.,

(i) The system (8.3) with $\omega(t) = 0$ is asymptotically stable;
(ii) Under the zero initial condition, $\|z(t)\|_2 < \gamma \|\omega(t)\|_2$ for any nonzero $\omega(t) \in L_2[0, \infty)$ and a prescribed $\gamma > 0$.

8.2 A Self-triggered Sampling Scheme

In this section, a self-triggered sampling scheme is proposed to reduce the use of limited network resources while maintaining the desired performance of the system under consideration.

8.2.1 An Upper Bound Estimation of an Error-Dependent Item

Define $\rho(t) \triangleq x(t) - x(t_k)$, $t \in [f_k, f_{k+1})$ as an error at the current sampling instant and the latest successfully transmitted sampling instant. We now state and establish the following stability criterion for the system (8.3).

Theorem 8.1 *For a given scalar $\gamma > 0$ and a matrix K, the closed-loop system (8.3) is asymptotically stable with an H_∞ disturbance attenuation level γ, if there exist matrices $M > 0$, $S > 0$, $W > 0$, $Q > 0$, $N > 0$, and $P > 0$ with appropriate dimensions such that*

(i) *for $t \in [f_k, f_{k+1})$,*

$$\rho^T(t)\Psi_1\rho(t) \leq x^T(t_k)\Psi_2 x(t_k), \tag{8.4}$$

where $\Psi_1 = M + S - W + N > 0$ and $\Psi_2 = W - WN^{-1}W - K^T D^T DK > 0$;

(ii)

$$\begin{bmatrix} \Omega & PB_1 \\ B_1^T P & \gamma^2 I \end{bmatrix} > 0 \tag{8.5}$$

$$\begin{bmatrix} Q - W & PBK & C^T DK \\ K^T B^T P & M & 0 \\ K^T D^T C & 0 & S \end{bmatrix} > 0 \tag{8.6}$$

where $\Omega = -P(A + BK) - (A + BK)^T P - C^T C - Q - C^T DK - K^T D^T C.$

Proof By Schur complement, (8.5) and (8.6) are written as

$$P\bar{A} + \bar{A}^T P + C^T C + \bar{B} + C^T DK + K^T D^T C < -Q \tag{8.7}$$

$$Q - \bar{P} - \bar{C} > W \tag{8.8}$$

where $\bar{A} = A + BK$, $\bar{P} = PBKM^{-1}K^T B^T P$, $\bar{B} = \gamma^{-2} PB_1 B_1^T P$, $\bar{C} = C^T DKS^{-1}K^T D^T C$.

Choose a Lyapunov function $V(x(t)) = x^T(t)Px(t)$, where $P > 0$. Notice that there exist matrices $M > 0$ and $S > 0$, and a scalar $\gamma > 0$ such that

$$x^T \bar{P}x(t) + \rho^T(t)M\rho(t) \geq -2x^T(t)PBK\rho(t) \tag{8.9}$$

$$x^T \bar{C}x(t) + \rho^T(t)S\rho(t) \geq -2x^T(t)C^T DK\rho(t) \tag{8.10}$$

$$x^T \bar{B}x(t) + \gamma^2 \omega^T(t)\omega(t) \geq 2x^T(t)PB_1\omega(t) \tag{8.11}$$

Taking the derivative of $V(x(t))$ with respect to t along the trajectory of (8.3) yields

$$\begin{aligned}
\dot{V}(x(t)) &= 2x^T(t)P(Ax(t) + BK x(t_k) + B_1\omega(t)) \\
&= x^T(t)(PA + A^T P)x(t) + 2x^T(t)PBK x(t_k) \\
&\quad + 2x^T(t)PB_1\omega(t) - z^T(t)z(t) \\
&\quad + (Cx(t) + DK x(t_k))^T(Cx(t) + DK x(t_k)) \\
&= x^T(t)(PA + A^T P + C^T C)x(t) - z^T(t)z(t) \\
&\quad + 2x^T(t)(PBK + C^T DK)(x(t) - \rho(t)) \\
&\quad + 2x^T(t)PB_1\omega(t) + x^T(t_k)K^T D^T DK x(t_k)
\end{aligned} \tag{8.12}$$

From (8.8)–(8.12), we have that

$$\begin{aligned}
\dot{V}(x(t)) &< \rho^T(t)(M + S)\rho(t) - x^T(t)Wx(t) - z^T(t)z(t) \\
&\quad + \gamma^2 \omega^T(t)\omega(t) + x^T(t_k)K^T D^T DK x(t_k)
\end{aligned} \tag{8.13}$$

From (8.4), for $N > 0$, we obtain that

$$
\begin{aligned}
\mathcal{M} &\le (x(t_k) + \rho(t))^T W(x(t_k) + \rho(t)) \\
&\quad -x^T(t_k)K^T D^T D K x(t_k) - 2x^T(t_k)W\rho(t) \\
&\quad -\rho^T(t)N\rho(t) - x^T(t_k)W N^{-1} W x(t_k) \\
&\le x^T(t)Wx(t) - x^T(t_k)K^T D^T D K x(t_k)
\end{aligned}
\tag{8.14}
$$

where $\mathcal{M} = \rho^T(t)(M+S)\rho(t)$. From (8.13) and (8.14), we have that

$$
\dot{V}(x(t)) < \gamma^2 \omega^T(t)\omega(t) - z^T(t)z(t)
\tag{8.15}
$$

Integrating both sides of (8.15) from f_k to t, $t \in [f_k, f_{k+1})$ and letting $k \to \infty$ and under zero initial condition, we obtain that $\|z(t)\|_2 < \gamma \|\omega(t)\|_2$. Moreover, when $\omega(t) = 0$, from (8.4) and (8.13), we have that $\dot{V}(x(t)) < -\lambda \|x(t)\|^2$ for a sufficiently small $\lambda > 0$. Then we can conclude that the system (8.3) is asymptotically stable. This completes the proof.   □

Notice that if $\rho(t)$ in (8.4) can be real-time computed by making use of the real-time measured state $x(t)$, and the number of successive packet dropouts satisfies a prescribed condition, an event-triggered sampling scheme, such as those proposed [69] can be used to determine the next sampling. However, since some existing event-triggered sampling schemes depend on specialized hardware for continuous measurement [64, 70, 78], therefore, it is infeasible to use the violation of the inequality in (8.4) to trigger the sampling or the control in most general devices.

To improve the energy efficiency for the system (8.1) controlled over a wireless network, in what follows, a self-triggered sampling scheme is proposed to estimate the next necessary sampling instant. The following theorem is important to estimate the upper bound of $\rho^T(t)\Psi_1\rho(t)$ in (8.4).

Theorem 8.2 *Consider the closed-loop system (8.3). The following inequality holds for $t \in [f_k, f_{k+1})$*

$$
\rho^T(t)\Psi_1\rho(t) \le \lambda_0(t, f_k, \omega(t), x(t_k))
\tag{8.16}
$$

where Ψ_1 is given in Theorem 8.1,

$$
\lambda_0(t, f_k, \omega(t), x(t_k)) = 2\lambda_1(t, f_k, \omega(t)) + 2\lambda_2^T(t, f_k, x(t_k))\lambda_2(t, f_k, x(t_k))
$$

and

$$
\lambda_1(t, f_k, \omega(t)) = (t - f_k)\int_{f_k}^t \|F_0\|^2 w^T(s)w(s)ds
$$

$$
\lambda_2(t, f_k, x(t_k)) = \int_{f_k}^t e^{F_2(s - f_k)}ds F_1 + e^{F_2(t - f_k)}\Psi_1^{\frac{1}{2}}\rho(f_k)
$$

with

$$F_0 = e^{F_2(t-s)} \Psi_1^{\frac{1}{2}} B_1, \; F_1 = \Psi_1^{\frac{1}{2}} (A + BK) x(t_k)$$
$$F_2 = \Psi_1^{\frac{1}{2}} A \Psi_1^{-\frac{1}{2}} \tag{8.17}$$

Proof Based on the system (8.3), we estimate the upper bound of $\rho^T(t)\Psi_1\rho(t)$ for $t \in [f_k, f_{k+1})$. Similar to [78], let $\Phi = \{t \in [f_k, f_{k+1}) : \rho^T(t)\Psi_1\rho(t) = 0\}$. The time derivative of $\Psi_1^{\frac{1}{2}} \rho(t)$ for $t \in [f_k, f_{k+1})\backslash\Phi$ satisfies

$$\frac{\mathrm{d}}{\mathrm{d}t} \Psi_1^{\frac{1}{2}} \rho(t) = \Psi_1^{\frac{1}{2}} (A(\rho(t) + x(t_k)) + BK x(t_k) + B_1\omega(t))$$
$$= F_1 + F_2 \Psi_1^{\frac{1}{2}} \rho(t) + \Psi_1^{\frac{1}{2}} B_1\omega(t) \tag{8.18}$$

where the right-hand side derivative is used when $t = f_k$ and F_1 and F_2 are defined in (8.17).

From (8.18) with the online measured initial condition $\Psi_1^{\frac{1}{2}} \rho(f_k)$, we have that

$$\Psi_1^{\frac{1}{2}} \rho(t) = \underbrace{\int_{f_k}^t e^{F_2(t-s)} \Psi_1^{\frac{1}{2}} B_1 w(s) \mathrm{d}s}_{\lambda_3(t, f_k, \omega(t))}$$

$$+ \underbrace{\int_{f_k}^t e^{F_2(s-f_k)} \mathrm{d}s F_1 + e^{F_2(t-f_k)} \Psi_1^{\frac{1}{2}} \rho(f_k)}_{\lambda_2(t, f_k, x(t_k))} \tag{8.19}$$

For simplicity, $\lambda_3(t, f_k, \omega(t))$ and $\lambda_2(t, f_k, x(t_k))$ in (8.19) are abbreviated as λ_3 and λ_2. Then from (8.19), we have that

$$[\Psi_1^{\frac{1}{2}} \rho(t)]^T \Psi_1^{\frac{1}{2}} \rho(t) = [\lambda_3 + \lambda_2]^T [\lambda_3 + \lambda_2]$$
$$= \lambda_3^T \lambda_3 + \lambda_2^T \lambda_2 + \lambda_3^T \lambda_2 + \lambda_2^T \lambda_3$$
$$\leq 2\lambda_3^T \lambda_3 + 2\lambda_2^T \lambda_2 \tag{8.20}$$

Using Lemma 2.1 in Chap. 2, for $t \in [f_k, f_{k+1})$, it is clear that

$$\lambda_3^T \lambda_3 = \int_{f_k}^t w^T(s) F_0^T \mathrm{d}s \int_{f_k}^t F_0 w(s) \mathrm{d}s$$
$$\leq (t - f_k) \int_{f_k}^t w^T(s) F_0^T F_0 w(s) \mathrm{d}s$$
$$\leq (t - f_k) \underbrace{\int_{f_k}^t \|F_0\|^2 w^T(s) w(s) \mathrm{d}s}_{\lambda_1(t, f_k, \omega(t))} \tag{8.21}$$

Then, from (8.20) and (8.21), we arrive at (8.16). This completes the proof. $\square$

In (8.16), the value of $x(t_k)$ is known a priori due to the fact that $x(t_k)$ is the latest successfully transmitted sampling value with the time-stamped t_k labeled by the sensor. Therefore, the case of data dropouts in the communication does not affect the utilization of Theorem 8.2. Furthermore, it is difficult to directly calculate $\lambda_1(t, f_k, \omega(t))$ since an unknown external disturbance $\omega(t)$ is involved in (8.16). However, since $\omega(t)$ belongs to an L_2 space, one can set $\|\omega(t)\|_2$ to its upper bound to make it solvable.

8.2.2 Self-triggered Sampling Scheme

Notice that Theorems 8.1 and 8.2 cannot be directly used to determine the next sampling instant since the nonideal network QoS, such as network-induced delays and data dropouts, inevitably exist in NCSs. To tackle the above problem, in this section, a self-triggered sampling scheme is developed to ensure the asymptotic stability of the system (8.3) with the desired performance by simultaneously taking network-induced delays and data dropouts into account.

The self-triggered sampling algorithm is developed with a diagram shown in Fig. 8.2.

Algorithm 1 Obtaining the sampling instant b_k for given τ_m, $\bar{w}$, τ_M, d_M and L

Fig. 8.2 A self-triggered sampling algorithm

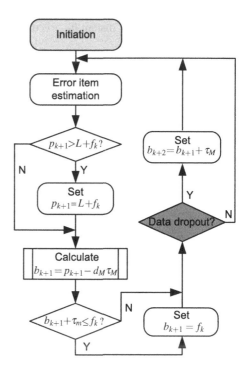

Step 1.   At f_k, the self-triggered sensor estimates the next violated instant p_{k+1}, which is the first value of t from a certain time violating the following inequality

$$\lambda_0(t, f_k, \bar{w}, x(t_k)) \leq x^T(t_k) \Psi_2 x(t_k) \qquad (8.22)$$

where $\bar{w}$ is the upper bound of $\omega(t)$. If $p_{k+1} > L + f_k$, then set $p_{k+1} = L + f_k$, where L is the prescribed allowable maximum sampling interval time between two consecutive samplings;

Step 2.   The next possible sampling instant b_{k+1} is set as $b_{k+1} = p_{k+1} - d_M \tau_M$, where d_M is the maximal allowable number of consecutive data dropouts, and τ_M is the upper bound of the network-induced delay τ_{t_k};

Step 3.   If $b_{k+1} + \tau_m < f_k$, then set $b_{k+1} = f_k$;

Step 4.   If the sampled packet with time-stamping t_{k+1} does not arrive at the actuator at $t \in [b_{k+1}, b_{k+1} + \tau_M)$, then the next sampling instant is set as $b_{k+2} = b_{k+1} + \tau_M$.

From Algorithm 1, one can see that the proposed self-triggered sampling scheme can tolerate a certain degree of nonideal QoS and make the proposed sampling scheme feasible. We now provide an explanation for Steps 3 and 4.

Step 3 is used to to deal with an unreasonable case in the self-triggered sampling. Figure 8.3 shows that $b_{k+1} < f_k$, where the predicted sampling instant p_{k+1} is determined by (8.22), and the computed sampling instant b_{k+1} is calculated based on Step 2. This implies that b_{k+1} is executed before f_k. In other words, the time order of next sampling f_k and the predicted sampling instant b_{k+1} is disorder. Thus, this predicted sampling instant b_{k+1} is discarded by Step 3.

Fig. 8.3 A sampling scheme with an invalid predicted sampling instant

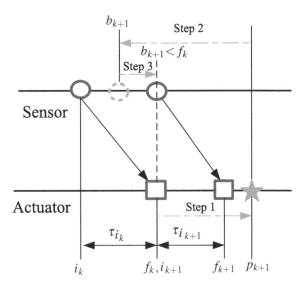

Step 4 is used to handle the case of data dropouts in the transmission. If there is no data dropout in the transmission, the control packet with the time-stamping b_{k+1} should arrive at the actuator before $b_{k+1} + \tau_M$, which is utilized by the self-triggered sensor to determine whether or not to sample the data. For example, when the packet arrived at the actuator at $t \in [b_{k+1}, b_{k+1} + \tau_M)$, the self-triggered sensor is triggered to compute p_{k+2}; in reverse, the next sampling instant is set as $b_{k+1} + \tau_M$ directly.

From the above description, one can see that the proposed self-triggered sampling scheme can be implemented in practice under consideration of the network-induced delay and data dropouts in the communications. Moreover, the maximal allowable successive number of data dropouts can be calculated as

$$d_M = \left\lceil \frac{p_{k+1} - f_k}{\tau_M} \right\rceil - 1 \tag{8.23}$$

where $\lceil \varkappa \rceil$ is the largest integer smaller than or equal to $\varkappa$.

Notice that the above-mentioned self-triggered sampling scheme includes cases of no data dropout and constant network-induced delays in the communications as its special cases. For example, when there is no data dropout in the communications, which means that $d_M = 0$, one can derive that $b_{k+1} = p_{k+1}$ in Step 2. In this case, the proposed self-triggered sampling scheme is still available.

If Step 3 is not included in Algorithm 1, it is possible that the next estimated instant $b_{k+1} + \tau_m$ arriving at the actuator is less than the instant f_k, that is, $f_k > b_{k+1} + \tau_m$. Clearly, it is not reasonable and should be avoided. Therefore, Step 3 has the role to ensure the feasibility of the computed sampling instant.

From Steps 1 and 3, it is derived that the largest and the least sampling instants are $L + f_k$ and f_k, respectively. Then, the least sampling interval can be calculated as

$$\min_k T_k = \min_k (b_{k+1} - b_k) = \tau_m, \tag{8.24}$$

where τ_m is the lower bound of the network-induced delay. Since τ_m is strictly larger than zero, it is concluded that $T_k > 0$. This is important due to the fact that there is no Zeno behavior [63] in the proposed scheme.

8.3 Stability Analysis of Self-triggered Sampling Scheme

By Theorems 8.1 and 8.2 and under the self-triggered sampling Algorithm 1, we have the following theorem for ensuring the asymptotic stability with an H_∞ disturbance attenuation level γ of the system (8.3).

Theorem 8.3 *For a given scalar $\gamma > 0$ and a matrix K, under the self-triggered sampling Algorithm 1, the system (8.3) is asymptotically stable with an H_∞ disturbance attenuation level γ and a maximal allowable number of consecutive data dropouts d_M; if there exist matrices $M > 0$, $S > 0$, $W > 0$, $Q > 0$, $N > 0$,*

and $P > 0$ with appropriate dimensions such that (8.5), (8.6), and (8.22) hold for
$t \in [f_k, f_{k+1})$.

Proof From Algorithm 1, it is known that the maximal allowable number of consecutive data dropouts is d_M. Furthermore, from (8.16) and (8.22), we have

$$\rho^T(t)\Psi_1\rho(t) \leq x^T(t_k)\Psi_2 x(t_k) \tag{8.25}$$

Then by Theorem 8.1, one can conclude that the system (8.3) is asymptotically stable
with an H_∞ disturbance attenuation level γ. This completes the proof.   □

The role of Theorem 8.2 is to estimate the upper bound of $\rho^T(t)\Psi_1\rho(t)$ in (8.4)
of Theorem 8.1, then this estimated value can be used in Theorem 8.3 to compare
with the value of $x^T(t_k)\Psi_2 x(t_k)$ in (8.22) to determine whether the current sampling
data should to be transmitted or not. Therefore, the advantage of Theorem 8.3 is that
some specialized hardware for continuous measurement and calculation of the value
of $\rho^T(t)\Psi_1\rho(t)$ in Theorem 8.1 is no longer needed.

8.4 An Example

In this section, an inverted pendulum controlled over a wireless network is used to
show the effectiveness of the proposed self-triggered sampling scheme. The plant's
state equation is given as [79, 125]

$$\dot{x}(t) = \begin{bmatrix} 0 & 1 \\ \frac{3(M+m)g}{l(4M+m)} & 0 \end{bmatrix} x(t) + \begin{bmatrix} 0 \\ -\frac{3}{l(4M+m)} \end{bmatrix} u(t) + B_1\omega(t) \tag{8.26}$$

where $M = 8.0$, $m = 2.0$, $l = 0.5$, $g = 9.8$, $B_1 = [1\ 0.1]^T$, $\omega(t) = 0.01\,|\sin(t)|$,
and $u(t) = [106.5970, 33.8599]\,x(t_k)$ as that [88].

Set $L = 0.36$ s and $\gamma = 200$. First, we consider two cases: The constant network-
induced delay $\tau_{t_k} = 0.09$ s, and the time-varying network-induced delay on an interval, $\tau_{t_k} \in [0.03\,\text{s}, 0.09\,\text{s}]$. With the different allowable number of successive data
dropouts, that is, $d_M = 0, 1, 2, 3$, the predicted sampling instants and the practically
used sampling instants based on the proposed self-triggered sampling scheme are
shown in Figs. 8.4 and 8.5, respectively, where "+" is the predicted sampling instant,
'◇' is the used sampling instant. From Figs. 8.4 and 8.5, one can see that:

(i) Most of the samplings occur at the system's adjustment period, such as $t \in$
[0, 10 s] in Figs. 8.4 and 8.5, and the sampling interval has lower and upper
bounds;

(ii) When the system state approaches the operating point, the self-triggered sampling scheme proposed in this paper has the lower sampling frequency while
maintaining the desired performance. Such as after 10 s, larger sampling intervals are generated in Figs. 8.4 and 8.5; and

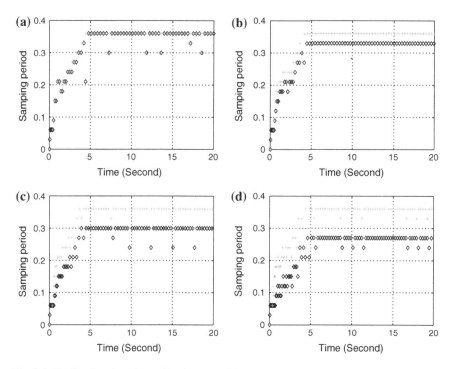

Fig. 8.4 Predicted and used sampling instants with a constant delay and different d_M. **a** $d_M = 0$, **b** $d_M = 1$, **c** $d_M = 2$, **d** $d_M = 3$

(iii) With the increase of the allowable number of uninterrupted data dropouts, the allowable sampling intervals decrease. This shows that the proposed self-triggered sampling scheme can adjust the sampling interval based on the the allowable number of data dropouts.

From (8.24), one can derive that the sampling interval should vary between 0.12 and 0.36 s in Fig. 8.4, and between 0.06 and 0.36 s in Fig. 8.5. However, it can be seen that there are some special instants, such as those in 0–5 s, are lower than 0.12 s. As mentioned in the paragraph around (8.24), the allowable lower bound of sampling period is used to deal with these special cases. In addition, with the given initial condition $x(0) = [2, -1]^T$, the system state responses are shown in Fig. 8.6.

For the purpose of comparison, Table 8.1 lists the sampled-data number in different sampling schemes for a simulation time $T = 30$ s. For the same number of allowable successive data dropouts, such as $d_M = 0, 1, 2$ and 3, respectively, one can see that the sampled-data number of the proposed self-triggered sampling scheme is smaller than that in the traditional time-triggered sampling scheme.

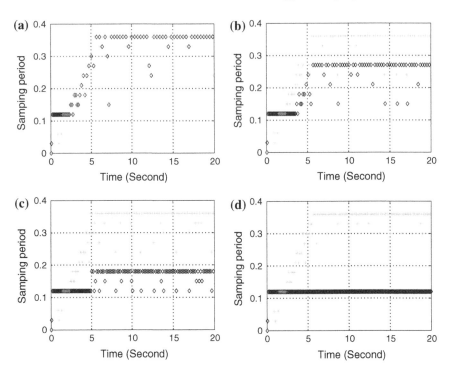

Fig. 8.5 Predicted and used sampling instants with time-varying delay and different d_M. **a** $d_M = 0$, **b** $d_M = 1$, **c** $d_M = 2$, **d** $d_M = 3$

Fig. 8.6 State responses of the closed-loop system with $d_M = 2$ in Fig. 8.4

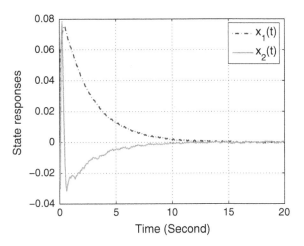

Table 8.1 Sampled-data number for different sampling schemes

Sampling schemes	$d_M = 0$	$d_M = 1$	$d_M = 2$	$d_M = 3$
Time-triggered in [88]	200	400	600	800
Self-triggered (varying delay)	108	138	189	252
Self-triggered (constant delay)	99	106	118	135

8.5 Conclusion

A self-triggered sampling scheme has been proposed for a wireless NCS by considering the network-induced delays and data dropouts simultaneously. This scheme plays a role to reduce the use of the limited communication bandwidth and energy, while ensuring the desired performance for the system under consideration. By using the error estimation technique, we have estimated the state value of the next sampling instant without depending on continuous measurement and online estimation of the event-triggered condition. Furthermore, the proposed sampling scheme can adaptively adjust the sampling interval based on the allowable number of successive data dropouts and the network-induced delays. A numerical example has been given to show the effectiveness of the proposed sampling scheme.

Chapter 9
A Mixed Sampling Scheme for Wireless Networked Control Systems

This chapter proposes a mixed self and event-triggered sampling scheme (MSE) for the execution of sampling in wireless networked control systems (WiNCSs) by striking a balance between self-triggered sampling (SET) and periodic event-triggered sampling (PET) to achieve a high energy efficiency. SET in MSE is used to reduction of idle listening of WiNCSs to save energy, and PET in MSE is used to reduce the conservativeness of SET by triggering the sampling events in a periodic manner. Compared with some existing ones, MSE does not require continuous measurement of the system state and does not suffer from the conservativeness induced by a self-triggering estimation. Moreover, PET in MSE is practically feasible and avoids the time mismatch problem. By using the proposed MSE in WiNCSs, one can: (i) Ensure the solution of the system under consideration is uniformly ultimately bounded (UUB); (ii) improve the energy efficiency in energy-constrained WiNCSs by reducing the number of transmitted packets; and (iii) ensure that there are lower and upper bounds of sampling interval, thereby avoiding the Zeno behavior and infinity sampling.

This chapter is organized as follows. The system modeling and problem description are provided in Sect. 9.1. Section 9.2 presents a MSE for the system under consideration controlled over wireless sensor networks, this section also provides an algorithm to obtain the next sampling instant determined by the proposed MSE. Stability analysis of NCSs under MSE is given in Sect. 9.3. A numerical example is given in Sect. 9.4. Finally, Sect. 9.5 concludes the chapter.

9.1 System and Problem Description

Consider the system described by

$$\dot{x}(t) = Ax(t) + Bu(t) + d(t) \tag{9.1}$$

© Springer-Verlag Berlin Heidelberg 2015
C. Peng et al., *Communication and Control for Networked Complex Systems*,
DOI 10.1007/978-3-662-46813-5_9

where $x(t) \in \mathbb{R}^n$ and $u(t) \in \mathbb{R}^m$ are the state vector and the control input vector, respectively; A and B are constant matrices with appropriate dimensions and $\|A\| \neq 0$; $d(t) \in \mathbb{R}^n$ is an exogenous bounded non-measurable disturbance, i.e., $\|d(t)\| \leq \bar{d}$. The initial condition of the system (9.1) is given by $x(t_0) = x_0$.

For the convenience, the following notations are introduced:

- t_k: The instant when the sensor gets the sampled-data from the plant and transmits the sampled-data to the controller.
- τ_k: The network-induced delay from the sensor to the actuator, which is the time from the instant t_k to the instant when the actuator accepted the time-stamped control packet.
- $\hat{t}_{k+1}$: The predicted sampling instant by a self-triggered estimator based on t_k.
- $f_k \triangleq t_k + \tau_k$: The instant when the time-stamped sampled packet arrived at the actuator.

For the sake of simplicity, in this chapter, we do not consider the the transmission delay in the controller-to-actuator channel, and assume that the communication delay τ_k introduced by the WSNs is time-varying and has the lower bound τ_m and upper bound τ_M. This implies that $0 < \tau_m \leq \tau_k \leq \tau_M$.

Define

$$e_k(t) \triangleq x_k - x(t), \quad t \in [f_k, f_{k+1}) \tag{9.2}$$

where x_k denotes $x(t_k)$, $e_k(t)$ is the error between two states at the latest transmitted sampling instant and the current instant.

When $t \in [f_k, f_{k+1})$, a zero-order holder (ZOH) between sensor and controller is used to keep the actuator input. Then the state feedback-based $u(t)$ can be written as

$$u(t^+) = Kx(t_k), \quad t \in [f_k, f_{k+1}) \tag{9.3}$$

where $u(t^+) = \lim_{\hat{t} \to t+0} u(\hat{t})$ and K is the state feedback controller gain.

From (9.1)–(9.3), we have the closed-loop system as

$$\dot{x}(t) = (A + BK)x(t) + BKe_k(t) + d(t), \quad t \in [f_k, f_{k+1}) \tag{9.4}$$

Since there is a WSN between the sensor and the controller, and the transmission operations consume most of the energy available at the nodes, an admissible sampling scheme should minimize the number of transmissions over the wireless networks to improve the energy efficiency [126, 127]. In the sequel, we are in a position to design an efficient sampling scheme to use as little as possible the limited energy while ensuring that the solution of the system (9.4) is UUB. These can be stated in the following problem definition.

Problem 9.1 Consider the system (9.4) and there is a WSN between the sensor and the controller, we aim at

(i) developing a SET with respect to external disturbance $d(t)$ to predict the $\hat{t}_{k+1}$ to extend as large as possible the inactive listening period of WSNs to reduce the energy expenditure while avoiding the Zeno behavior and infinity sampling.

(ii) proposing a step-state-error-dependent PET to determine whether or not the current sampled-data should to be transmitted to save limited network bandwidth and energy while avoiding the time mismatch between the threshold condition of SET and PET;

(iii) constructing a MSE, which combines the schemes given in (i) and (ii) in a unified framework to improve the energy efficiency while ensuring the desired control performance.

To tackle Problem 9.1, in the following section, a SET is developed to ensure that the solution of the system (9.4) is UUB; then based on the predicted sampling instant by SET, a step-state-error-dependent PET is proposed to reduce the conservativeness of the inter-sampling times induced by SET; Finally, a MSE is provided, which makes best use of the advantages of SET and PET and couples them in a unified framework.

9.2 A Mixed Sampling Scheme

The conceptual diagram of WiNCSs with a MSE is shown in Fig. 9.1, where solid lines represent physical links and broken lines for the next sampling and communication decision flows; the controller and the actuator are event-triggered and the sampled-data of the plant is transmitted with a single packet. In contrast to a typical NCS structure with the event-triggered [63, 128] or the self-triggered sampling schemes [75, 78], the key characteristic of the conceptual diagram proposed in this chapter is that a MSE sampling scheme is introduced in Fig. 9.1.

Moreover, MSE is composed by a SET and a PET, where

- SET is used to predict the next sampling instant, which begins to estimate the next sampling instant $\hat{t}_{k+1}$ when the sampled-data arrived at the controller.
- PET is located following the SET. From the instant $\hat{t}_{k+1}$, PET begins to check the predesigned event-triggered conditions periodically to determine t_{k+1}.

Figure 9.2 shows a case of how to determine t_{k+1} in the proposed MSE. For example, when $t \in [t_{k-1} + \tau_{k-1}, t_k + \tau_k)$, one can see from Fig. 9.2 that the sampled packet at the predicted instant $\hat{t}_k$ is not transmitted, and the event-triggered sampler online periodically check the event-triggered conditions and then to determine that the sampled-data at the instant $t_k = \hat{t}_k + 3h$ should be transmitted though the wireless sensor networks.

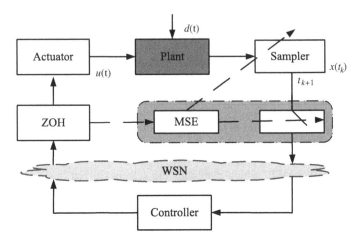

Fig. 9.1 A framework of WiNCSs with a MSE

Fig. 9.2 An example of time evolution of a MSE communication scheme. "★" means the predicted sampling instant by SET; "■" means the instant of the controller packet arrived at the ZOH; and "○" means the checking instant of the event-triggered sampler with the checking period h

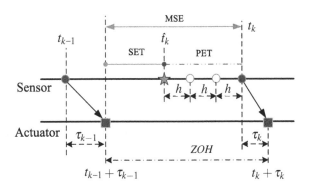

9.2.1 A Self-triggered Estimator

In this section, a lemma is first given to estimate the error between the state at the current instant and the state at the latest transmitted sampling instant. Then, a feasible self-triggered estimator is developed to generate the next sampling instant while ensuring that the solution of the system (9.4) is UUB.

Different from the reactive fashion in an event-triggered sampling scheme [59], the self-triggered sampling [74] is performed in a predictive fashion. Since the predictive fashion can reduce the idle listening period while reducing the energy expenditure, in what follows, similar to [64, 75], the following SET is used as part of the proposed MSE to generate the next sampling for the system under consideration. Lemma 9.1 and Theorem 9.1 can be derived by following similar way used in [64, 75]. For completeness, they are also listed in this chapter.

Before developing a SET, the following lemma is necessary to evaluate the upper bound of the measurement error $\|e_k\|$ on the system (9.4).

Lemma 9.1 *Consider the system* (9.4), *for* $t \in [f_k, f_{k+1})$, *the upper bound of the error* $\|e_k\|$ *is*

$$\|e_k\| \le g(t, x_{k-1}, x_k, \bar{d}, \tau_k) \tag{9.5}$$

where

$$
\begin{aligned}
g(t, &x_{k-1}, x_k, \bar{d}, \tau_k) \\
&:= \|A\|^{-1} \exp(\|A\| (t - t_k - \tau_k))(\|A x_k + B K x_{k-1}\| + \bar{d})(\exp(\|A\| \tau_k) - 1) \\
&+ \|A\|^{-1} (\|(A + BK)x_k\| + \bar{d})(\exp(\|A\| (t - t_k - \tau_k) - 1)
\end{aligned} \tag{9.6}
$$

Proof From the definition of $e_k := x_k - x(t), t \in [f_k, f_{k+1})$, we have $\dot{e}_k = -\dot{x}(t)$. Then for $t \in (f_k, f_{k+1})$, we have

$$
\begin{aligned}
\|\dot{e}_k\| &= \|-\dot{x}(t)\| = \|A(e_k - x_k) - B K x_k - d(t)\| \\
&\le \|A\| \|e_k\| + \|(A + BK)x_k\| + \bar{d}
\end{aligned} \tag{9.7}
$$

By making use of the Comparison Lemma [129], the upper bound of $\|e_k\|$ can be written as

$$
\begin{aligned}
\|e_k\| &\le \exp(\|A\| (t - f_k)) \|e_k(f_k)\| \\
&+ \frac{\|(A + BK)x_k\| + \bar{d}}{\|A\|} \times (\exp(\|A\| (t - f_k)) - 1)
\end{aligned} \tag{9.8}
$$

To estimate $\|e_k(f_k)\|$ in (9.8), similar to the operation in (9.7) and (9.8), for $t \in (t_k, f_k)$, we have

$$
\begin{aligned}
\|\dot{e}_k\| &= \|-\dot{x}(t)\| = \|A(e_k - x_k) - B K x_{k-1} - d(t)\| \\
&\le \|A\| \|e_k\| + \|A x_k + B K x_{k-1}\| + \bar{d}
\end{aligned} \tag{9.9}
$$

$$
\begin{aligned}
\|e_k\| &\le \exp(\|A\| (t - t_k)) \|e_k(t_k)\| \\
&+ \frac{\|A x_k + B K x_{k-1}\| + \bar{d}}{\|A\|} \times (\exp(\|A\| (t - t_k)) - 1)
\end{aligned} \tag{9.10}
$$

Due to the fact that at the sampling instants $t = t_k, e_k(t_k) = 0$, from (9.10), we have

$$\|e_k(f_k)\| \le \frac{\|A x_k + B K x_{k-1}\| + \bar{d}}{\|A\|} \times (\exp(\|A\| (\tau_k)) - 1) \tag{9.11}$$

Combining (9.8) and (9.11), we arrive at (9.5). This completes the proof. □

The idea of SET is to predict the time it takes for $\|e_k\|$ to go from $\|e_k(f_k)\|$ to δ, i.e.,

$$\max_t \{t \mid \|e_k(t)\| \leq \delta, \quad t \in [f_k, f_{k+1})\} \tag{9.12}$$

where $\delta > 0$ is a pre-given scalar. In this way, we can bound the sampling error $\|e_k\|$ through δ to estimate the next maximum allowable sampling instant $\hat{t}_{k+1}$.

Based on Lemma 9.1, the following UUB criterion is established.

Theorem 9.1 *For given $\bar{d} > 0$, $\tau_M > 0$, $\theta \in (0, 1)$ and δ satisfying*

$$\delta > \mu =: \|A\|^{-1} (\|Ax_k + BKx_{k-1}\| + \bar{d})(\exp(\|A\| \tau_M) - 1) \tag{9.13}$$

with the following self-triggered estimator

$$\hat{t}_{k+1} = t_k + \gamma(x_{k-1}, x_k, \bar{d}, \tau_k) \tag{9.14}$$

the solution of the closed-loop system (9.4) is UUB with a ultimate bound

$$\|x\| \leq \frac{2\lambda_{\max}(P)(\|BK\|\delta + \bar{d})}{\theta c} \sqrt{\frac{\lambda_{\max}(P)}{\lambda_{\min}(P)}} := b \tag{9.15}$$

where

$$\gamma(x_{k-1}, x_k, \bar{d}, \tau_k) := \frac{1}{\|A\|} \ln \left(\frac{\Psi(x_k, \bar{d})}{\Xi(x_k, x_{k-1}, \bar{d}, \tau_k)} \right) + \tau_k \tag{9.16}$$

with

$$\Psi(x_k, \bar{d}) = \|A\|\delta + \|(A + BK)x_k\| + \bar{d} \tag{9.17}$$

$$\Xi(x_k, x_{k-1}, \bar{d}, \tau_k) = (\|Ax_k + BKx_{k-1}\| + \bar{d})(\exp(\|A\| \tau_k) - 1)$$
$$+ \|(A + BK)x_k\| + \bar{d} \tag{9.18}$$

Proof First, we prove that the predicted value of $\hat{t}_{k+1}$ is given in (9.14). From the mechanism of the proposed self-triggered estimator described in (9.5) and (9.12), the next maximum sampling instant $\hat{t}_{k+1}$ is

$$\hat{t}_{k+1} = t_k + \max_t \{t \mid g(t, x_{k-1}, x_k, \bar{d}, \tau_k) \leq \delta\} \tag{9.19}$$

Based on Lemma 9.1, one can obtain the maximum value of t satisfying the constraint in (9.5) equals to $\gamma(x_{k-1}, x_k, \bar{d}, \tau_k)$ in (9.16). Then (9.14) follows. Moreover, for making the self-triggered estimator reasonable, $\hat{t}_{k+1}$ should be larger than f_k, we can derive that $\Psi(x_k, \bar{d}) > \Xi(x_k, x_{k-1}, \bar{d}, \tau_k)$ from (9.16), then (9.13) is needed.

Second, we prove that the the solution of the closed-loop system (9.4) is UUB by using the self-triggered estimator (9.14). Consider the Lyapunov candidate $V(x) = x^T P x$, where $P = P^T > 0$. Let $P(A+BK)^T + (A+BK)P = -Q, a_1 = \lambda_{\min}(P)$, $a_2 = \lambda_{\max}(P)$ and $c = \lambda_{\min}(Q)$. For $t \in [f_k, f_{k+1})$, the derivative of the Lyapunov candidate function along the trajectories of (9.4) satisfies

$$
\begin{aligned}
\dot{V}(x) &= -x^T Q x + 2x^T P(BKe_k + d(t)) \\
&\leq -c \|x\|^2 + 2x^T P(BKe_k + d(t))
\end{aligned}
\tag{9.20}
$$

by exploiting the continuity of $V(x)$ and considering sampling scheme (9.14), we can further upper bound $\dot{V}(x)$ as

$$
\begin{aligned}
\dot{V}(x) &\leq -c \|x\|^2 + 2a_2 \|x\| (\|BK\| \delta + \bar{d}) \\
&\leq -(1-\theta)c \|x\|^2, \quad \forall \|x\| \geq \frac{2a_2(\|BK\| \delta + \bar{d})}{\theta c}
\end{aligned}
\tag{9.21}
$$

where $0 < \theta < 1$.

For the chosen Lyapunov candidate, it holds that $a_1 \|x\|^2 \leq V(x) \leq a_2 \|x\|^2$. By using the similar operation in [129] (Chap. 4, Theorem 4.18), one can readily derive that the solution of the system (9.4) is UUB with a ultimate bound defined in (9.15). This completes the proof. $\qquad\square$

In practical engineering application, no matter what kind of sampling scheme is established, it is required that there is no Zeno behavior [63], i.e., the minimum sampling interval $h_m := \min\{h := \hat{t}_{k+1} - f_k\} > 0$. However, Theorem 9.1 does not provide any information about the minimum sampling interval. The following Corollary shows that the predicted sampling periods generated by the proposed self-triggered sampling scheme (9.14) are bounded away from zero.

Corollary 9.1 *The minimum inter-event time h_m of the self-triggered estimator (9.14) is lower bounded by*

$$
h_m \geq F(x_k, x_{k-1}, \bar{d}, \tau_k) \big|_{\|x_k\| = \|x_{k-1}\| = M(k, \delta, t_0), \tau_k = \tau_M}
\tag{9.22}
$$

specially, after the transient process

$$
\begin{aligned}
F(x_k, x_{k-1}, \bar{d}, \tau_k) \big|_{\|x_k\| = \|x_{k-1}\| = 0, \tau_k = \tau_m} &\geq h_m \\
\geq F(x_k, x_{k-1}, \bar{d}, \tau_k) \big|_{\|x_k\| = \|x_{k-1}\| = b, \tau_k = \tau_M}
\end{aligned}
\tag{9.23}
$$

where $\Xi(x_k, x_{k-1}, \bar{d}, \tau_k)$ being defined in (9.18), b being defined in (9.15) and

$$F(x_k, x_{k-1}, \bar{d}, \tau_k) := \frac{1}{\|A\|} \ln\left(\frac{\Lambda(x_k, x_{k-1}, \bar{d}, \tau_k)}{\Xi(x_k, x_{k-1}, \bar{d}, \tau_k)} + 1\right)$$

$$\Lambda(x_k, x_{k-1}, \bar{d}, \tau_k) := \|A\|\,\delta - (\|Ax_k + BKx_{k-1}\| + \|\bar{d}\|)$$
$$\times (\exp(\|A\|\,\tau_k) - 1)$$

$$M(k, \delta, t_0) := \exp(\|A + BK\|\,\tau_M)(k\delta + \|x(t_0)\|)$$
$$+ \frac{\|BK\|\,\delta + \bar{d}}{\|A + BK\|} \times (\exp(\|A\|\,\tau_M) - 1)$$

Proof From (9.14), we have

$$\begin{aligned}
h_m &:= \min\{\hat{t}_{k+1} - f_k\} \\
&\geq \min\{\frac{1}{\|A\|} \ln\left(\frac{\Psi(x_k, \bar{d})}{\Xi(x_k, x_{k-1}, \bar{d}, \tau_k)}\right)\} \\
&= \min\{\frac{1}{\|A\|} \ln\left(\frac{\Lambda(x_k, x_{k-1}, \bar{d}, \tau_k)}{\Xi(x_k, x_{k-1}, \bar{d}, \tau_k)} + 1\right)\}
\end{aligned} \tag{9.24}$$

where $\Lambda(x_k, x_{k-1}, \bar{d}, \tau_k)$ being given in Corollary 9.1.

From (9.13), it is readily to show that $\Lambda(x_k, x_{k-1}, \bar{d}, \tau_k) > 0$ and $\frac{\Lambda(x_k, x_{k-1}, \bar{d}, \tau_k)}{\Xi(x_k, x_{k-1}, \bar{d}, \tau_k)} > 0$ in (9.24). Then one can see that h_m is monotonically decreasing on the decision variables x_k, x_{k-1} and τ_k.

For $t \in (f_k, f_{k+1})$, from (9.4) and (9.12), we have

$$\|\dot{x}(t)\| \leq \|A + BK\|\,\|x(t)\| + \|BK\|\,\delta + \bar{d} \tag{9.25}$$

For $t \in [t_k, f_k)$, using the Comparison Lemma [129], we have

$$\begin{aligned}
\|x(f_k)\| &\leq \exp(\|A + BK\|\,\tau_k)\,\|x(t_k)\| \\
&+ \frac{\|BK\|\,\delta + \bar{d}}{\|A + BK\|} \times (\exp(\|A\|\,\tau_k) - 1)
\end{aligned} \tag{9.26}$$

From (9.12), for $t \in (f_k, f_{k+1})$, it is clear that

$$\|x(t_{k+1}) - x(t_k)\| \leq \delta \tag{9.27}$$

From (9.27), we have

$$\|x(t_k)\| \leq \delta + \|x(t_{k-1})\| \leq k\delta + \|x(t_0)\| \tag{9.28}$$

From (9.26) and (9.28), we have

$$\|x(f_k)\| \leq \exp(\|A + BK\| \tau_M)(k\delta + \|x(t_0)\|)$$
$$+ \frac{\|BK\| \delta + \bar{d}}{\|A + BK\|} \times (\exp(\|A\| \tau_M) - 1) \qquad (9.29)$$

Therefore, we can set $\|x_k\|$ and $\|x_{k-1}\|$ as $M(k, \delta, t_0)$, and τ_M as τ_k to get the h_m in (9.22). Specially, after the transient process, the solution of the closed-loop system (9.4) is UUB with ultimate bound (9.15), that is $\|x_k\| \leq b$, then from (9.24), we can set $\|x_k\|$ and $\|x_{k-1}\|$ as b and τ_M as τ_k to achieve (9.23). This completes the proof. □

From (9.23), one can see that the predicted sampling instants have lower and upper bounds. This implies that the proposed SET sampling scheme is away from zero sampling and infinity sampling.

For the measured variables x_{k-1}, x_k, τ_k and t_k, from the self-trigged sampling scheme (9.14), one can see that the predictive next sampling instant $\hat{t}_{k+1}$ is monotonically decreasing on the decision variable $\bar{d}$ over the domain specified by the constraint (9.13), especially, when $\|e_k(\hat{t}_{k+1})\|$ is far away from δ. Therefore, making use of upper bound $\bar{d}$ to replace $d(t)$ in (9.14) is conservative to estimate the next sampling instant $\hat{t}_{k+1}$. As a result, large number of unnecessary sampled packets are transmitted over the networks resulting in low energy efficiency. In what follows, before proposing a novel MSE for a WiNCS to reduce the conservativeness induced by SET, the following backstepping PET sampling scheme is necessary.

9.2.2 A Backstepping Periodic Event-Triggered Sampling Scheme

From (9.12), one can see that if $\|e_k(t)\|$ can be online computed based on real-time measured state $x(t)$, the following event-triggered sampling scheme

$$t_{k+1} = t_k + \min\{t \,|\|e_k(t)\| \geq \delta\} \qquad (9.30)$$

can be directly used to trigger the sampling [59]. However, an event-triggered sampler based on (9.30) depends on specialized hardware for continuous measurement [64, 70]. Therefore, using the violation of the inequality in (9.30) to trigger the sampling or the control is impractical in most general devices [130].

For avoiding the continuous measurement, a PET has been proposed in [67, 69, 128]. The idea of PET can be described as

$$t_{k+1} = f_k + \min_{l \in \mathbb{N}}\{lh \,|\|e_k(f_k + lh)\| \geq \delta\} \qquad (9.31)$$

Fig. 9.3 An example of the communication instant with different sampling schemes (★ predicted instant by SET; △ the value of $\delta - \lambda(l)$; ○ used instant by MSE; ◇ the instant of $\|e(t_2 + h)\| > \delta$)

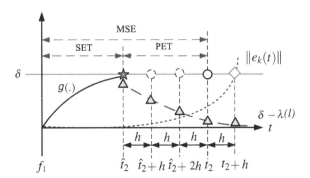

where $e_k(f_k + lh) := x(t_k + lh) - x(t_k)$, $l = 1, 2, \ldots$, h is the checking period. Compared with continuous measurement and calculation of the value of $\|e_k(t)\|$ in (9.30), one can see that specialized hardware for continuous measurement and calculation is no longer needed in (9.31). However, PET (9.31) cannot be directly applied in this chapter, since at the triggered instant t_{k+1} determined by (9.31), it is possible that $\|e_k(t_{k+1})\| > \delta$ while deviating from the required condition (9.12). This is illustrated by Fig. 9.3. One can see from Fig. 9.3 that at the instant t_2, $\|e(t_2)\| < \delta$, and at the instant $t_2 + h$, $\|e(t_2 + h)\| > \delta$. Based on (9.31), the triggered instant is $t_2 + h$. However, the ideal transmitted instant satisfying (9.12) should be t_2. Due to the measured periodicity of (9.31), only at the instant $t_2 + h$ can we find this fact. As a result, the condition (9.12) is no longer satisfied at the practically transmitted instant $t_2 + h$. In other words, there is a time mismatch problem while directly using PET [67, 69, 128] to follow SET.

Based on the above analysis, the following event-triggered sampling scheme is proposed to guarantee that at the transmitted instant t_{k+1} the condition (9.12) is satisfied.

$$t_{k+1} = f_k + \max_{l \in \mathbb{N}^0}\{lh \mid \|e_k(f_k + lh)\| \le \delta\} \tag{9.32}$$

However, since the delay notified property is included in (9.32), that is, only at instant $(l+1)h$ we can know that lh satisfies (9.32). Therefore, based on (9.32), it is difficult to directly find the maximum l and trigger the communication simultaneously. To make the triggering condition (9.32) available in practical application, in what follows, an estimate-based PET is proposed. For the development, the following lemma is necessary to evaluate the upper bound of step-state error $\|\Delta x(t)\|$, where $\|\Delta x(t)\|$ is defined as

$$\Delta x(t) = x(t + h) - x(t) \tag{9.33}$$

Lemma 9.2 *For a given sampling step $h > 0$, the upper bound of the step-state error from the instant $\hat{t}_{k+1} + lh$ to the instant $\hat{t}_{k+1} + (l + 1)h, l = 0, 1, 2, \ldots$ is*

$$\left\| \Delta x(\hat{t}_{k+1} + lh) \right\| \le \lambda(l, x_k, x_{\hat{t}_{k+1}}, h, \bar{d}) \tag{9.34}$$

where

$$\lambda(l, x_k, x_{\hat{t}_{k+1}}, h, \bar{d}) := \exp(\|A\| \, lh) \Omega(x_k, x_{\hat{t}_{k+1}}, h, \bar{d})$$

$$+ \frac{2\bar{d}}{\|A\|} \times (\exp(\|A\| \, lh) - 1) \tag{9.35}$$

$$\Omega(x_k, x_{\hat{t}_{k+1}}, h, \bar{d}) := (\exp(\|A\| \, h) + 1) \, \|x(\hat{t}_{k+1})\|$$

$$+ \frac{\|(A + BK)x_k\| + \bar{d}}{\|A\|} \times (\exp(\|A\| \, h) - 1) \tag{9.36}$$

Proof From (9.4) and (9.33), it is clear

$$\|\Delta\dot{x}(t)\| = \|A\Delta x(t) + d(t+h) - d(t)\| \le \|A\| \, \|\Delta x(t)\| + 2\bar{d} \tag{9.37}$$

Using the Comparison Lemma [129], for $t = \hat{t}_{k+1} + lh, l = 0, 1, 2, \ldots$, we have

$$\|\Delta x(t)\| \le \exp(\|A\| \, (t - \hat{t}_{k+1})) \, \|\Delta x(\hat{t}_{k+1})\|$$

$$+ \frac{2\|\bar{d}\|}{\|A\|} \times (\exp(\|A\| \, (t - \hat{t}_{k+1})) - 1) \tag{9.38}$$

From (9.33), it is derived that

$$\|\Delta x(\hat{t}_{k+1})\| \le \|x(\hat{t}_{k+1} + h)\| + \|x(\hat{t}_{k+1})\| \tag{9.39}$$

To the known $\|x(\hat{t}_{k+1})\|$ at the instant $\hat{t}_{k+1}$, from Lemma 9.1, we have

$$\|x(\hat{t}_{k+1} + h)\| \le \frac{\|(A + BK)x_k\| + \bar{d}}{\|A\|} (\exp(\|A\| \, h) - 1)$$

$$+ \exp(\|A\| \, h) \, \|x(\hat{t}_{k+1})\| \tag{9.40}$$

Combining (9.38)–(9.40), inequality (9.34) follows. This completes the proof. □

In the sequel, for notational simplify, $\lambda(l)$ denotes $\lambda(l, x_k, x_{\hat{t}_{k+1}}, h, \bar{d})$. Based on Lemma 9.2, we propose the following backstepping PET scheme

$$t_{k+1} = f_k + \min_{l \in \mathcal{N}^0} \{ lh \, | \, \|e_k(f_k + lh)\| \ge \delta - \lambda(l) > 0 \} \tag{9.41}$$

Remark 9.1 The function of $\lambda(l)$ in (9.41) is to ensure the condition (9.12), i.e., $\|e(\hat{t}_{k+1} + lh)\| < \delta$ at the discrete communication instant $\hat{t}_{k+1} + lh$, where lh is determined by (9.41). Since the event-triggered condition included in (9.41) can be real-time measured to determine lh, which overcomes the drawback of unmeasure condition given in (9.32).

Remark 9.2 Notice that if l equals to 0 in (9.41), the proposed backstepping PET is simplified as an NCS with an event-triggered sensor, that is, the event of sampled-data arrived at the controller triggers the sensor to sample and transmit the sampled-data simultaneously.

From (9.41), we have the following corollary, which shows that the proposed backstepping PET guarantees the event-triggering condition (9.12) at the triggered instant $t \in [f_k, f_{k+1})$.

Corollary 9.2 *With the given step-state-error-dependent PET (9.41), the following inequality is guaranteed,*

$$\delta \geq \|e_k(f_k + lh)\| \geq \delta - \lambda(l) > 0 \tag{9.42}$$

where l is the value online determined by (9.41).

Proof Based on PET sampling scheme described in (9.41), when $l \neq 0$, we know that if l is the minimum value satisfying the condition in (9.41):

$$\|e_k(f_k + (l-1)h)\| < \delta - \lambda(l-1) \tag{9.43}$$

Moreover, it is clear

$$
\begin{aligned}
\|e_k(f_k + lh)\| \ &\leq \|e_k(f_k + (l-1)h)\| \\
&\quad + \|e_k(f_k + lh) - e_k(f_k + (l-1)h)\|
\end{aligned}
\tag{9.44}
$$

From Lemma 9.2, it is known that

$$
\begin{aligned}
\|e_k(f_k + lh) &- e_k(f_k + (l-1)h)\| \\
&= \|x(f_k + lh) - x(f_k + (l-1)h)\| \leq \lambda(l-1)
\end{aligned}
\tag{9.45}
$$

Combining (9.43), (9.44) and (9.45), we have

$$\|e_k(f_k + lh)\| \leq \|e_k(f_k + (l-1)h)\| + \lambda(l-1) < \delta \tag{9.46}$$

Then from (9.41) and (9.46), we arrive at (9.42); When $l = 0$, from the triggered scheme (9.41), we have

$$\|e_k(f_k)\| \geq \delta - \lambda(0) > 0 \tag{9.47}$$

Moreover, based on the self-triggered condition (9.12), at instant f_k:

$$\delta \geq \|e_k(f_k)\| \tag{9.48}$$

We also have (9.42) from (9.47) and (9.48). This completes the proof.   □

From Theorem 9.1 and Corollary 9.1, one can see that the self-triggered estimator (9.14) can be used to generate the next sampling time with lower and upper bounds while guaranteeing that the solution of the system (9.4) is UUB. However, similar to [127], since we use a worst-case disturbance acting over all the time in the derivations of Lemma 9.1 and Theorem 9.1, resulting in the next sampling instants generated by (9.14) are too conservative if there are no disturbances acting on the process [75]. Moreover, the event-triggered sampling does not contribute to the reduction of the idle listening because nodes are enforced to keep the radio on for all the time to wait for the transmission of the event-generated data. To reduce the above-mentioned conservativeness while saving the limited bandwidth and energy, in the next section, we will show how to solve the points (iii) in Problem 9.1.

9.2.3 A Mixed Self and Event-Triggered Sampling Scheme

From (9.14) and (9.41), we present the following mixed self and event-triggered sampling scheme

$$t_{k+1} = \hat{t}_{k+1} + \min_{l \in \mathbb{N}^0} \{lh \mid \|e(\hat{t}_{k+1} + lh)\| \geq \delta - \lambda(l) > 0\} \qquad (9.49)$$

where δ satisfying (9.13), $\hat{t}_{k+1}$ being given in (9.14) and $\lambda(l)$ being defined in (9.34), respectively.

The running process of the proposed MSE sampling scheme is summarized as Algorithm 2.

Algorithm 2 Find the next sampling instant t_{k+1} based on the MSE.

Step 1. At the instant t_k, use the self-triggered sampling scheme (9.14) to estimate the $\hat{t}_{k+1}$;

Step 2. Compute $\lambda(l)$ based on (9.34) for $l = 0$. If $\delta - \lambda(l) > 0$ and $\|e(\hat{t}_{k+1} + lh)\|$ $< \delta - \lambda(l)$, set $l = l + 1$ and compute $\lambda(l)$ based on (9.34); and

Step 3. Set the next sampling time as $t_{k+1} = \hat{t}_{k+1} + lh$.

Remark 9.3 Notice that Step 2 in Algorithm 2 is to find the minimum value of l based on the triggering condition given in (9.49). As a special case, when $l = 0$, that is, in the case of $\delta - \lambda(0) \leq \|e(\hat{t}_{k+1})\| < \delta$, MSE is simplified as a SET described in (9.14). Moreover, if $\hat{t}_{k+1}$ in (9.49) is replaced by t_k, then MSE would be simplified as PET (9.41). Furthermore, different from PET (9.41), when $l = 0$, the requirement of $e(\hat{t}_{k+1})\| < \delta$ at the minimum transmitted instant is satisfied by MSE (9.49).

Remark 9.4 Notice that the constraint of $\delta - \lambda(l) > 0$ provides a method to tradeoff l and h in (9.41). From the definition of $\lambda(l, x_k, x_{\hat{t}_{k+1}}, h, \bar{d})$ in (9.35), it is known that the value of $\delta - \lambda(l)$ is decreasing with the increase of l and h. For reducing the conservativeness of inter-event times, in this chapter, we set $h = h_m$ determined by (9.23).

Remark 9.5 It is worth mentioning that the sampling intervals determined by (9.49) have nonzero lower and upper bounds. It is because that: (i) From (9.16), we know $\hat{t}_{k+1} - t_k > \tau_k > 0$; and (ii) for the known x_k, $x_{\hat{t}_{k+1}}$ and $\bar{d}$ in (9.35), one can see that the function $\lambda(l, x_k, x_{\hat{t}_{k+1}}, h, \bar{d})$ is monotonically increasing on the decision variables h and l. These issues imply that the l satisfying the constraint in (9.49) is upper bounded and away from infinity. Therefore, for the given h, the value of $\delta - \lambda(l)$ in (9.49) is decreasing on the l. This shows that $t_{k+1} - t_k$ is upper bounded.

9.3 Stability Analysis of NCSs Under Mixed Self and Event-Triggered Sampling Scheme

Now we are in a position to show that under the proposed MSE (9.49), the solution of the closed-loop system (9.4) is UUB.

Theorem 9.2 *For given $\bar{d} > 0$, $\tau_M > 0$, $\theta \in (0, 1)$ and $\delta > \mu$, μ being defined in (9.13), by using the mixed self and event-triggered sampling scheme (9.49), the solution of the closed-loop system (9.4) is UUB with a ultimate bound b given in (9.15).*

Proof Under the mixed sampling rule (9.49), we divide $t \in [f_k, f_{k+1})$ as $t \in [f_k, \hat{t}_{k+1})$, $t \in [\hat{t}_{k+1}, t_{k+1})$ and $t \in [t_{k+1}, f_{k+1})$. For $t \in [f_k, \hat{t}_{k+1})$, based on the self-triggered estimator, it is clear $\|e_k(t)\| \leq \delta, t \in [f_k, \hat{t}_{k+1})$; for $t \in [\hat{t}_{k+1}, t_{k+1})$, from the periodic event-triggered sampling property described in (9.41), we can guarantee that $\|e_k(t)\| \leq \delta, t \in [\hat{t}_{k+1}, t_{k+1})$; and for $t \in [t_{k+1}, f_{k+1})$, from Lemma 9.1, it is derived that

$$\|e_k(t)\| \leq \frac{\|Ax_k + BKx_{k-1}\| + \bar{d}}{\|A\|} \times (\exp(\|A\|(\tau_k)) - 1) \qquad (9.50)$$

Combining (9.13) and (9.50), we have $\|e_k(t)\| \leq \delta, t \in [t_{k+1}, f_{k+1})$. Based on above description, one can see that the condition $\|e_k\| \leq \delta, t \in [f_k, f_{k+1})$ is kept based on (9.49). Then the precondition in the derivation of Theorem 9.1 is satisfied, then the remainder proof is similar the proof of Theorem 9.1, it is omitted here. This completes the proof.   □

From Theorem 9.2, it is known that the control performance is guaranteed with the proposed MSE. Compared with the application of SET and PET alone, the highest energy efficient can be expected based on MSE. This will be verified by the following Section.

9.4 An Example

In this section, we use an example to show the effectiveness of the proposed MSE.

An inverted pendulum is controlled over a WSN. The plant's state equation is [117]

$$\dot{x}(t) = \begin{bmatrix} 0 & 1 \\ \frac{3(M+m)g}{l(4M+m)} & 0 \end{bmatrix} x(t) + \begin{bmatrix} 0 \\ -\frac{3}{l(4M+m)} \end{bmatrix} u(t) + d(t) \tag{9.51}$$

where $M = 8.0$, $m = 2.0$, $l = 0.5$, $g = 9.8$, and $u(t)$ as given in (9.3) with $K = [106.5970, 33.8599]$ [88].

Assume $d(t) = [\sin(2\pi t); \sin(2\pi t)]$ and the network-induced delay is time-varying in a known range, i.e., $\tau_k \in [0.01\,\text{s}, 0.05\,\text{s}]$, and also set that the event-triggered sampling period $h = 0.02\,\text{s}$, $\delta = 0.19$. The release times and the system trajectories are shown in Figs. 9.4, 9.5 and 9.6 by making use of SET, PET, and MSE, respectively. From Figs. 9.4 and 9.6, one can see:

(i) In the system's transient process, the release times of SET and MSE are very approximately. However, after the adjustment period, such as after 7 s in the first subfigure of Fig. 9.6, larger communication intervals are generated in MSE than those in SET/PET. This verifies the role of event-triggered component in MSE. For example, from the second subfigure of Fig. 9.6, one can see that when $t \in [0, 7\,\text{s}]$, the role of event-triggered component to increase the transmission interval is slight; however, when $t > 7\,\text{s}$, the event-triggered component in MSE has a notable role to increase the transmission interval; and

(ii) the conservativeness induced by SET/PET is reduced by making use of the proposed MSE. For example, compared with the number of transmitted packets 243 in SET and 403 in PET, only 149 samplings are transmitted in MSE, and

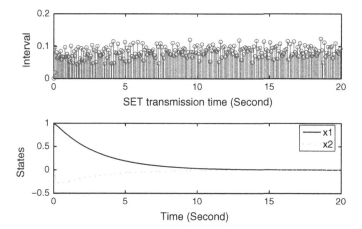

Fig. 9.4 Communication condition, state trajectories of SET

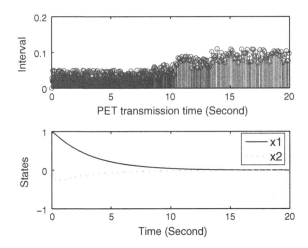

Fig. 9.5 Communication condition, state trajectories of PET

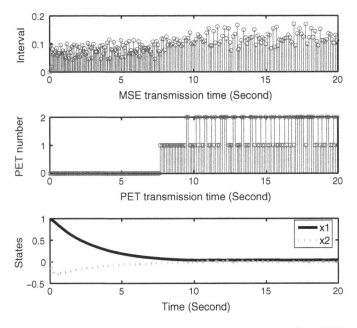

Fig. 9.6 Communication condition, Event-triggered packets, State trajectories of MSE

63 % of triggered packets in PET is no longer needed to be transmitted in MSE. This responses that MSE can decrease the number of transmitted packets to save the limited energy of WSN.

Moreover, for comparisons, based on Algorithm 2, Table 9.1 lists the number of transmitted packets among different sampling schemes for a simulation time $T = 20$ s

Table 9.1 Number of transmitted packets for different sampling schemes

Sampling methods	$h = 0.02$	$h = 0.04$	$h = 0.06$	$h = 0.08$
Time-triggered [89]	1000	500	334	250
SET [79]	243	243	243	243
PET [128]	403	270	221	168
MSE of this chapter	149	145	148	144
Improvement over PET (%)	63.0	46.3	33.3	14.3
Improvement over SET (%)	38.7	40.3	39.1	40.1

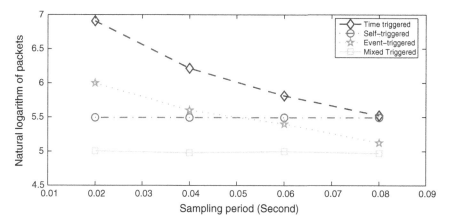

Fig. 9.7 Natural logarithm of the number of transmitted packets with different sampling periods and sampling schemes

with event-triggered sampling periods $h = 0.02, 0.04, 0.06$, and 0.08 s. One can see that the number of transmitted packets of the proposed MSE is the least with different allowable sampling periods. For example, with $h = 0.04$ s, compared with the time-triggered sampling scheme in [88, 89], on average only 29 % sampled-data need to be transmitted; compared with the self-triggered sampling scheme in [79, 127], on average only 60 % sampled data need to be transmitted; compared with the periodic event-triggered sampling scheme in [128], on average only 53.7 % sampled-data need to be transmitted. In particular, after the adjustment time i.e., $t > 7$ s in Fig. 9.6, the MSE proposed in this chapter has the lower communication frequency while maintaining the desired performance. Figure 9.7 depicts the natural logarithm of the number of transmitted packets with different sampling periods and sampling schemes, which also verifies the above-mentioned advantages of the proposed MSE sampling scheme.

For comparisons, the following performance index is proposed to compare the energy efficiency of different sampling schemes.

Table 9.2 Consumed energy for different sampling schemes

Sampling methods	$h = 0.02$	$h = 0.04$	$h = 0.06$	$h = 0.08$
Time-triggered [89]	850	425	284	213
SET [79]	207	207	207	207
PET [128]	432	264	204	156
MSE of this chapter	179	146	136	131
Improvement over PET (%)	58.6	44.7	33.3	16.0
Improvement over SET (%)	13.5	29.5	34.3	36.7

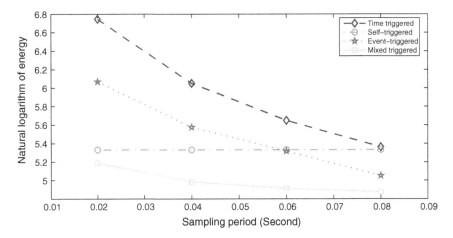

Fig. 9.8 Natural logarithm of consumed energy with different sampling periods and sampling schemes

$$J_i = \sum_{j=1}^{3} k_j^i N_j^i, \quad i = 1, 2, 3, 4 \tag{9.52}$$

J_i mean the consumed energy of the different sampling schemes (J_1:time-triggered; J_2:SET; J_3:PET; J_4:MSE); k_j^i are the weighting coefficients of different processes ($j = 1$: computation of SET; $j = 2$: sampling; $j = 3$: radio of triggered packets); N_j^i are the number of corresponding energy consumption operations of the different sampling schemes. In this chapter, set $k_1^i = k_2^i = 0.15$, and $k_3^i = 0.7$, since the radio operation consumes most of the limited energy of WSNs and the consumed energy by the computation of SET at every self-triggered instant and the periodic sampling by PET are similar.

Based on (9.52), the consumed energy with different sampling schemes and sampling periods are listed in Table 9.2. The natural logarithm of consumed energy with different sampling periods and sampling schemes is depicted in Fig. 9.8. This verifies that the proposed MSE sampling scheme has the highest energy efficiency than others listed in Table 9.2.

From the above simulation, one can see that the proposed MSE has a notable role to reduce the number of transmitted packets compared with SET and PET. Therefore, the wireless communication in a WiNCS based on MSE can achieve the highest energy efficiency than those based on SET and PET.

9.5 Conclusion

A mixed sampling scheme in a WiNCS has been proposed to ensure that the solution of the system under consideration is UUB, while improving the energy efficiency of the WSNs. By coupling the developed SET sampling and the proposed backstepping PET sampling techniques in a unified framework, we have obtained the highest energy efficiency than some existing ones; so the proposed MSE has fully used the advantage of SET and PET to extend the inactive period of the WSNs, and to reduce the conservativeness induced by SET. Moreover, the time mismatch problem in the application and SET and PET have been well solved. Two numerical examples have been given to show the effectiveness of the proposed method.

Chapter 10
Event-Triggered Control for Networked Takagi–Sugeno Fuzzy Systems

In this chapter, a discrete event-triggered communication scheme is used to save the limited network resource while preserving the desired performance and without resorting to extra hardware, where "discrete" means that we only measure the state and compute the error at a constant sampling period. The discrete measured state is used to determine whether or not the measured state should be transmitted. Compared with those continuous event-triggered schemes in [10, 59], where the extra hardware is necessary to fulfil the continuous measurements and computation [10, 16, 70], since we only measure the state and compute the error at a constant sampling period, the extra hardware for continuous measurement is not necessary. Different from those in [10] through simulations to show that the transmission periods are greater than a positive constant as the state goes to the equilibrium, our proposed discrete transmission scheme guarantees that the lower bound on the transmission period is a constant sampling period. Morevoer, to reduce the conservertiveness induced by the asynchronous premises in T-S fuzzy plant and PDC fuzzy rules under network environments, the method mentioned in Sect. 4.1.2 is also used to reconstruct the synchronous time scale grades of membership at the controller as those in the T-S fuzzy systems [51].

The rest of this chapter is organized as follows. Section 10.1 presents an event-triggered communication scheme for T-S fuzzy systems. Section 10.2 constructs a networked T-S fuzzy system model with asynchronous premise constraints, which has the advantage in reducing the conservativeness induced by asynchronous premises of T-S plant and the PDC fuzzy rules. Section 10.3 develops an event-triggered controller design method subject to asynchronous grades of membership. Two numerical examples are given in Sect. 10.4. Finally, Sect. 10.5 concludes the chapter.

© Springer-Verlag Berlin Heidelberg 2015
C. Peng et al., *Communication and Control for Networked Complex Systems*,
DOI 10.1007/978-3-662-46813-5_10

10.1 An Event-Triggered Communication Scheme

In this section, a discrete event-triggered communication scheme proposed in Chap. 2 is used to reduce the number of the transmitted packets while preserving the stability and desired control performance. The main idea is to transmit the measurements only when the variations of the system state at the current sampling exceed the specified threshold. That is, the system state is sampled at a constant sampling period h, whether or not the sampled state $x(jh)$, $j \in \mathbb{N}$, should be transmitted is determined by the transmission-error and the state error, i.e., the error between the current system state and the latest transmitted state, and the error between the current system state and its equilibrium.

For ease of exposition, the following assumptions are needed in this chapter:

- Sensors are clock-driven. The system states are sampled at a constant period h. The set of sampled instants is represented by $\{jh \,|\, j \in \mathbb{N}\}$;
- Controllers and actuators are event-driven. The control packets carry the time stamps of the transmitted packets. The logic zero-order holder (ZOH) is used to hold the control input, when there is no the latest control packet arrived at the actuator [110];
- The transmitted instant $t_k h$ is determined by the sampled state $x(jh)$. The set of transmission instants is represented by $\{t_k h \,|\,\{t_k \in \mathbb{N}\}$. All transmitted packets are time-stamped;
- Network-induced delays from the sensor to the controller and from the controller to the actuator, and the computational and waiting delays are lumped together as τ_{t_k}, where $\tau_{t_k} \in (0, \bar{\tau}]$, $\bar{\tau}$ is the upper bound of τ_{t_k}.

From the above assumptions, one can see that the set of transmission instants $\{t_k h \,|\, t_k \in \mathbb{N}\}$ is a subset of $\{jh \,|\, j \in \mathbb{N}\}$, and the holding zone of ZOH is composed of the following subsets

$$\Omega = [t_k h + \tau_{t_k}, t_{k+1}h + \tau_{t_{k+1}}) = \cup_{n=0}^{d}\Omega_{n,k} \tag{10.1}$$

where

$$\Omega_{n,k} = [i_k h + \tau_{t_k}, i_k h + h + \tau_{t_{k+1}})$$
$$i_k h = t_k h + nh; n = 0, \ldots, d; d = t_{k+1} - t_k - 1 \tag{10.2}$$

and $i_k h$ means the sampling between the two conjoint transmitted instant; τ_{t_k} and $\tau_{t_{k+1}}$ are the network-induced delays at the transmitting instants $t_k h$ and $t_{k+1}h$, respectively.

Based on the above description, the state error between the current sampling instant and the latest transmission instant can be calculated as

$$e_k(i_k h) = x(i_k h) - x(t_k h) \tag{10.3}$$

Define $\eta(t) \triangleq t - i_k h$, $t \in \Omega_{n,k}$. From (10.3), the transmitted state $x(t_k h)$ can be written as

$$x(t_k h) = x(t - \eta(t)) - e_k(i_k h), \quad t \in \Omega_{n,k} \tag{10.4}$$

From the definition of $\eta(t)$, one can see that $\eta(t)$ is a differentiable function satisfying

$$\dot{\eta}(t) = 1, 0 < \tau_{t_k} \le \eta(t) \le h + \tau_{t_{k+1}} \le h + \bar{\tau} \triangleq \bar{\eta}, \quad t \in \Omega_{n,k} \tag{10.5}$$

where h and $\bar{\tau}$ are the sampling period and the allowable upper network-induced delay bound, respectively.

In this chapter, we assume that the occurrence of a transmission is dependent on an error-dependent discrete event rather than the passing of time, which decides when the next transmission should be taken place. The next transmission instant determined by the event generator given in Sect. 2.2.1 can be expressed as

$$t_{k+1}h = t_k h + \inf_{n \ge 0} \{nh \, \big| \, e_k^T(i_k h)\Phi e_k(i_k h) \ge \delta x^T(i_k h)\Phi x(i_k h)\} \tag{10.6}$$

where $\delta > 0$ is given scalar parameters, and Φ is a given positive definite weighting matrix. From (10.6), one can see that the transmission events are dependent not only on the state-dependent error $e_k(i_k h)$, but also on the current states $x(i_k h)$. When the condition listed in (10.6) is satisfied, the transmission events are triggered.

Compared with the time-triggered transmission scheme in [43, 44, 125], where all the sampled data are transmitted regardless of any situation of the system state, the used event-triggered scheme in this chapter does not transmit the sampled state except that it violates from the preselected threshold, and thus can reduce the number of packets transmitted through the communication network. Compared with continuous event-triggered transmission scheme in [10, 59], where the special hardware is necessary for successive measurement and computation, the used event-triggered scheme in this chapter is a discrete event-triggered transmission scheme because it is only dependent on the latest transmitted state $\{t_k h \, | t_k \in \mathbb{N}\}$ and state error at the discrete sampling sets $\{jh \, | j \in \mathbb{N}\}$. Therefore, the special hardware for continuous measurement and calculation mentioned in [10] is no longer needed.

10.2 T-S Fuzzy Systems with Asynchronous Premise Constraints

Consider a T-S fuzzy system, where ith rule of the system is expressed in the following If-Then rule

$$R^i: \text{If } \theta_1(t) \text{ is } W_1^i \text{ and}, \dots, \text{ and } \theta_g(t) \text{ is } W_g^i,$$

$$\text{Then } \begin{cases} \dot{x}(t) = A_i x(t) + B_i u(t) + B_{\omega i}\omega(t) \\ z(t) = C_i x(t) + D_i u(t) \end{cases} \tag{10.7}$$

where $i = 1, 2, \ldots, r$, r is the number of If-Then rules; $x(t) \in \mathbb{R}^n$ and $u(t) \in \mathbb{R}^m$ are the state vector and the input vector, respectively; W_j^i ($i = 1, 2, \ldots, r$; $j = 1, 2, \ldots, g$) are fuzzy sets; and $\theta_j(t)$ ($j = 1, 2, \ldots, g$) represent the premise variables. Denote $\theta(t) = [\theta_1(t), \ldots, \theta_g(t)]^T$, and assume that $\theta(t)$ is either given or a function of $x(t)$ and does not depend on $u(t)$. The input $\omega(t) \in L_2[0, \infty)$ denotes the exogenous disturbance signal; $z(t) \in \mathbb{R}^p$ represents the system output; the initial condition of the system (10.7) is given by $x(t_0) = x_0$; A_i, B_i, $B_{\omega i}$, C_i and D_i ($i = 1, 2, \ldots, r$) are constant matrices with compatible dimensions.

By using a center-average defuzzifier, product inference, and a singleton fuzzifier, the global dynamics of the T-S fuzzy system (10.7) can be inferred as

$$
\begin{cases}
\dot{x}(t) = \sum_{i=1}^{r} \mu_i(\theta(t))[A_i x(t) + B_i u(t) + B_{\omega i} \omega(t)] \\
z(t) = \sum_{i=1}^{r} \mu_i(\theta(t))[C_i x(t) + D_i u(t)]
\end{cases}
\tag{10.8}
$$

where $\mu_i(\theta(t)) = \frac{h_i(\theta(t))}{\sum\limits_{i=1}^{r} h_i(\theta(t))}$, $h_i(\theta(t)) = \Pi_{j=1}^{g} W_j^i(\theta_j(t))$, and $W_j^i(\theta_j(t))$ is the membership value of $\theta_j(t)$ in W_j^i.

Suppose that the system (10.7) is controlled over a communication network and the system state is available for feedback. In the following, we will design a T-S fuzzy model-based controller via a parallel distributed compensation (PDC) to stabilize the T-S fuzzy system (10.7). Since there exists a communication network between the sensor and the controller in an NCS, which implies that the available time stamped packet to derive the premises in the system and the controller should be asynchronous. That is, at the same instant $t \in [t_k h + \tau_{t_k}, t_{k+1} h + \tau_{t_{k+1}}) \triangleq \Omega$, although the premise variables $\theta_i(t)$ is available in (10.8), only $\theta_i(t_k h)$ is available at the controller. Based on the above description, the ith state feedback controller rule can be designed as

$$
\begin{aligned}
R^i : &\text{If } \theta_1(t_k h) \text{ is } W_1^i \text{ and }, \ldots, \text{ and } \theta_g(t_k h) \text{ is } W_g^i, \\
&\text{Then } u(t) = K_j x(t_k h), \quad t \in \Omega
\end{aligned}
\tag{10.9}
$$

where K_j, $j = 1, 2, \ldots, r$, are controller gains to be determined.

Combining (10.4) and (10.9) together, the defuzzified output of the PDC controller is

$$
u(t) = \sum_{j=1}^{r} \mu_j(\theta(t_k h)) K_j(x(t - \eta(t)) - e_k(i_k h)), \quad t \in \Omega_{n,k}
\tag{10.10}
$$

Notice that the premise variables $\mu_j(\theta(t_k h))$ of (10.10) are different from $\mu_i(\theta(t))$ of (10.8) for $t \in \Omega$. That is, the premise variables of T-S fuzzy system and PDC fuzzy rules are asynchronous.

To reduce the conservativeness by directly using (10.10) in the derivation of the main results; in the following, the premise reconstructed method mentioned in Chap. 4 [51] is also used. Assume that

$$\begin{cases} \mu_j^k = \rho_j \mu_j \\ |\mu_j^k - \mu_j| \le \Delta_j \end{cases} \tag{10.11}$$

where ρ_j, Δ_j $(j = 1, 2, \ldots, r)$ are some positive constants, μ_j and μ_j^k represent $\mu_j(\theta(t))$ and $\mu_j(\theta(t_k h))$, respectively.

Based on the asynchronous constraints (10.11), it is clear that

$$v_1^j = 1 - \frac{\Delta_j}{\mu_j} \le \rho_j \le 1 + \frac{\Delta_j}{\mu_j} = v_2^j \tag{10.12}$$

where v_1^j and v_2^j denote the minimum and maximum values of ρ_j during the operation, then, we have

$$\frac{v_1^i}{v_2^j} = \frac{\min\{\rho_i\}}{\max\{\rho_j\}} \le \frac{\rho_i}{\rho_j} \le \frac{\max\{\rho_i\}}{\min\{\rho_j\}} = \frac{v_2^i}{v_1^j} \tag{10.13}$$

Setting $v_1 = \min\{v_1^i\}$ and $v_2 = \max\{v_2^i\}$ $(i = 1, 2, \ldots, r)$, (10.13) yields that

$$\lambda_1 = \frac{v_1}{v_2} \le \frac{\rho_i}{\rho_j} \le \frac{v_2}{v_1} = \lambda_2 \tag{10.14}$$

From (10.11), (10.10) can be written as

$$u(t) = \sum_{j=1}^{r} \rho_j \mu_j K_j (x(t - \eta(t)) - e_k(i_k h)), \quad t \in \Omega_{n,k} \tag{10.15}$$

Substituting (10.15) into (10.8) leads to the following closed-loop fuzzy system

$$\dot{x}(t) = \mathscr{A}^i x(t) + \mathscr{B}_j^i x(t - \eta(t)) - \mathscr{B}_j^i e_k(i_k h) + \mathscr{B}_\omega^i \omega(t) \tag{10.16a}$$

$$z(t) = \mathscr{C}^i x(t) + \mathscr{D}_j^i x(t - \eta(t)) - \mathscr{D}_j^i e_k(i_k h), \text{ for } t \in \Omega_{n,k} \tag{10.16b}$$

where

$$\mathscr{A}^i = \sum_{i=1}^{r} \mu_i A_i, \, \mathscr{B}_\omega^i = \sum_{i=1}^{r} \mu_i B_{\omega i}, \, \mathscr{C}^i = \sum_{i=1}^{r} \mu_i C_i,$$

$$\mathscr{B}_j^i = \sum_{i=1}^{r} \sum_{j=1}^{r} \rho_j \mu_i \mu_j B_i K_j, \, \mathscr{D}_j^i = \sum_{i=1}^{r} \sum_{j=1}^{r} \rho_j \mu_i \mu_j D_i K_j.$$

We supplement the initial condition of the state $x(t)$ on $[t_0 - \bar{\eta}, t_0]$ as $x(t_0 + \theta) = \phi(\theta)$, $\theta \in [-\bar{\eta}, 0]$, with $\phi(0) = x_0$, where $\phi(\theta)$ is a continuous function on $[t_0 - \bar{\eta}, t_0]$.

Under the proposed event-triggered communication framework, the purpose of this chapter is to design a controller (10.10) such that:

(i) The system (10.16a) with $\omega(t) = 0$ is asymptotically stable;

(ii) Under the zero initial state condition, the H_∞ performance $\|z(t)\| < \gamma \|\omega(t)\|$ is guaranteed for any nonzero $\omega(t) \in L_2[0, \infty)$ and a prescribed $\gamma > 0$.

10.3 Stability Analysis and Controller Design

This section is to develop an approach for stability analysis and controller synthesis of the networked T-S fuzzy closed-loop system (10.16). The result is summarized in the following theorem. One can see that both the event-triggered communication scheme (10.6) and the inner network characteristics are fully utilized in the derivation of our result.

We now state and establish the following theorem.

Theorem 10.1 *For some given positive constants $\bar{\eta}$, γ, λ_1, λ_2 and matrix K_j, under the communication scheme (10.6), the system (10.16) is asymptotically stable with an H_∞ norm bound γ, if there exist real matrices $T > 0$, $P_i > 0$, $R_i > 0$ ($i = 1, 2$) and matrices M, N, and L with appropriate dimensions such that the following linear matrix inequalities (LMIs) hold for $k, i, j = 1, 2, \ldots, r$, $i < j$, $l = 1, 2$:*

$$\begin{bmatrix} \Xi_{11}^{kk}(l) & * \\ \Xi_{21}^{kk}(l) & \Xi_{22} \end{bmatrix} < 0 \tag{10.17}$$

$$\begin{bmatrix} \Xi_{11}^{ij}(l) + \lambda_1 \Xi_{11}^{ji}(l) & * & * \\ \Xi_{21}^{ij}(l) & \Xi_{22} & * \\ \sqrt{\lambda_2} \Xi_{21}^{ji}(l) & 0 & \Xi_{22} \end{bmatrix} < 0 \tag{10.18}$$

$$\begin{bmatrix} \Xi_{11}^{ij}(l) + \lambda_2 \Xi_{11}^{ji}(l) & * & * \\ \Xi_{21}^{ij}(l) & \Xi_{22} & * \\ \sqrt{\lambda_2} \Xi_{21}^{ji}(l) & 0 & \Xi_{22} \end{bmatrix} < 0 \tag{10.19}$$

where

$$\Xi_{11}^{ij}(l) = \Gamma_1^{ij} + \Delta + \Delta^T + (2 - l)\Gamma_2^{ij}$$
$$\Xi_{21}^{ij}(1) = col\{\bar{\eta} T F_2^{ij}, \bar{\eta} R_2 F_2^{ij}, \bar{\eta} L^T, F_3^{ij}\}$$
$$\Xi_{21}^{ij}(2) = col\{\bar{\eta} T F_2^{ij}, \bar{\eta} N^T, \bar{\eta} M^T, F_3^{ij}\}$$
$$\Xi_{22} = -diag\{\bar{\eta} T, \bar{\eta} R_2, \bar{\eta} T, I\}$$
$$\Gamma_2^{ij} = \bar{\eta} F_1^T R_1 F_2^{ij} + \bar{\eta}(F_2^{ij})^T R_1 F_1$$
$$\Delta = [M + N, -M + L, 0, -L, -N, 0]$$

$$F_1 = [I, 0, 0, 0, -I, 0]$$
$$F_2^{ij} = [A_i, B_i K_j, -B_i K_j, 0, 0, B_{\omega i}]$$
$$F_3^{ij} = [C_i, D_i K_j, -D_i K_j, 0, 0, 0]$$

and

$$\Gamma_1^{ij} = \begin{bmatrix} P_1 A_i + A_i^T P_1 + P_2 - R_1 & * & * & * & * & * \\ K_j^T B_i^T P_1 & \delta \Phi & * & * & * & * \\ -K_j^T B_i^T P_1 & 0 & -\Phi & * & * & * \\ 0 & 0 & 0 & -P_2 & * & * \\ R_1 & 0 & 0 & 0 & -R_1 & * \\ B_{\omega i}^T P_1 & 0 & 0 & 0 & 0 & -\gamma^2 \end{bmatrix}$$

Proof Construct a Lyapunov functional candidate as

$$V(t, x_t) = V_1(t, x_t) + V_2(t, x_t) \tag{10.20}$$

where

$$V_1(t, x_t) = x^T(t) P_1 x(t) + \int_{t-\bar{\eta}}^{t} x^T(v) P_2 x(v) dv + \int_{t-\bar{\eta}}^{t} \int_{s}^{t} \dot{x}^T(v) T \dot{x}(v) dv ds$$

$$V_2(t, x_t) = (\bar{\eta} - \eta(t))\{[x^T(t) - x^T(s_k)] R_1 [x(t) - x(s_k)] + \int_{s_k}^{t} \dot{x}^T(v) R_2 \dot{x}(v) dv\}$$

and $T > 0$, $P_i > 0$, and $R_i > 0$ $(i = 1, 2)$, $s_k = i_k h + \tau_{t_k}, t \in \Omega_{n,k}$.

Using the Newton-Leibnitz formula, for matrices M, N and L of appropriate dimensions, we have

$$2\rho^T(t) M[x(t) - x(t - \eta(t)) - \int_{t-\eta(t)}^{t} \dot{x}(s) ds] = 0$$

$$2\rho^T(t) L[x(t - \eta(t)) - x(t - \bar{\eta}) - \int_{t-\bar{\eta}}^{t-\eta(t)} \dot{x}(s) ds] = 0$$

$$2\rho^T(t) N[x(t) - x(s_k) - \int_{s_k}^{t} \dot{x}(s) ds] = 0$$

where $\rho_1^T(t) = [x^T(t), x^T(t - \eta(t)), e_k^T(i_k h), x^T(t - \bar{\eta}), x^T(s_k)]$ and $\rho^T(t) = [\rho_1^T(t), \omega^T(t)]$. For the sake of simplicity, in the following, $\rho(t)$ is denoted by ρ. Notice that there exist real matrices $R_2 > 0$ and $T > 0$ such that

$$-2\rho^T M \int_{t-\eta(t)}^t \dot{x}(s)ds \le \eta(t)\rho^T MT^{-1}M^T \rho + \int_{t-\eta(t)}^t \dot{x}^T(s)T\dot{x}(s)ds \quad (10.21)$$

$$-2\rho^T L \int_{t-\bar{\eta}}^{t-\eta(t)} \dot{x}(s)ds \le (\bar{\eta} - \eta(t))\rho^T LT^{-1}L^T \rho + \int_{t-\bar{\eta}}^{t-\eta(t)} \dot{x}^T(s)T\dot{x}(s)ds$$

$$\quad (10.22)$$

$$-2\rho^T N \int_{s_k}^t \dot{x}(s)ds \le (t - s_k)\rho^T NR_2^{-1}N^T \rho + \int_{s_k}^t \dot{x}^T(s)R_2\dot{x}(s)ds$$

$$\le \eta(t)\rho^T NR_2^{-1}N^T \rho + \int_{s_k}^t \dot{x}^T(s)R_2\dot{x}(s)ds \quad (10.23)$$

From the definition of $i_k h$, it is concluded that $i_k h \in [t_k h, t_{k+1}h - h]$. Then, from (10.6), it is known that the next transmitted instant is $t_{k+1}h$, which means that

$$e_k^T(i_k h)\Phi e_k(i_k h) < \delta x^T(i_k h)\Phi x(i_k h) \quad (10.24)$$

Taking the time derivative of (10.20) for system (10.16) and using $\frac{d}{dt}x(s_k) = 0$ and $\dot{\eta}(t) = 1$, for $t \ne i_k h + \tau_{t_k}$ and $t \in \Omega_{n,k}$, we have

$$\dot{V}(t, x_t) \le \sum_{i=1}^r \sum_{j=1}^r \mu_i(\theta(t))\mu_j(\theta(t_k h))\rho^T (\Xi_{11}^{ij}(l) - (\Xi_{21}^{ij}(l))^T \Xi_{22}^{-1}\Xi_{21}^{ij}(l))\rho$$
$$- z^T(t)z(t) + \gamma^2 \omega^T \omega(t) \quad (10.25)$$

where $\Xi_{11}^{ij}(l)$, $\Xi_{21}^{ij}(l)$, Ξ_{22} are defined in Theorem 10.1.

Since $\mu_j(\theta(t_k h)) = \rho_j \mu_j(\theta(t))$ adopted in (10.11), we have

$$\dot{V}(t, x_t) \le \sum_{i=1}^r \sum_{j=1}^r \rho_j \mu_i \mu_j \rho^T \left\{ \Xi_{11}^{ij}(l) - \Sigma_{ij}(l) \right\} \rho - z^T(t)z(t) + \gamma^2 \omega^T(t)\omega(t)$$

$$= \sum_{i=1}^{r-1} \sum_{j>i}^r \rho_j \mu_i \mu_j \rho^T \left\{ \Xi_{11}^{ij}(l) + \Xi_{11}^{ji}(l) - \Sigma_{ij}(l) - \frac{\rho_i}{\rho_j}\Sigma_{ji}(l) \right\} \rho$$

$$+ \sum_{i=1}^r \rho_i \mu_i^2 \rho^T(t) \left\{ \Xi_{11}^{ii}(l) - \Sigma_{ii}(l) \right\} \rho(t) - z^T(t)z(t) + \gamma^2 \omega^T(t)\omega(t)$$

$$\quad (10.26)$$

where $\Sigma_{ij}(l) = (\Xi_{21}^{ij}(l))^T \Xi_{22}^{-1}\Xi_{21}^{ij}(l) < 0$.

Using Schur Complement (Lemma 2.1 in Chap. 2), (10.17), (10.18) and (10.19) imply that

$$\Xi_{11}^{kk}(l) - \Sigma_{kk}(l) < 0 \tag{10.27}$$

$$\Xi_{11}^{ij}(l) + \lambda_1 \Xi_{11}^{ji}(l) - \Sigma_{ij}(l) - \lambda_2 \Sigma_{ji}(l) < 0 \tag{10.28}$$

$$\Xi_{11}^{ij}(l) + \lambda_2 \Xi_{11}^{ji}(l) - \Sigma_{ij}(l) - \lambda_2 \Sigma_{ji}(l) < 0 \tag{10.29}$$

Considering (10.28) and (10.29) with $(\lambda_2 - \frac{\rho_i}{\rho_j})\Sigma_{ji} < 0$, we have

$$\Xi_{11}^{ij}(l) + \lambda_1 \Xi_{11}^{ji}(l) - \Sigma_{ij}(l) - \frac{\rho_i}{\rho_j}\Sigma_{ji}(l) < 0 \tag{10.30}$$

$$\Xi_{11}^{ij}(l) + \lambda_2 \Xi_{11}^{ji}(l) - \Sigma_{ij}(l) - \frac{\rho_i}{\rho_j}\Sigma_{ji}(l) < 0 \tag{10.31}$$

Define $\varepsilon_1 = \frac{\lambda_2 - \frac{\rho_i}{\rho_j}}{\lambda_2 - \lambda_1} \geq 0$ and $\varepsilon_2 = \frac{\frac{\rho_i}{\rho_j} - \lambda_1}{\lambda_2 - \lambda_1} \geq 0$. It follows from (10.30) and (10.31) that

$$\sum_{m=0,1} \varepsilon_m (\Xi_{11}^{ij}(l) + \lambda_m \Xi_{11}^{ji}(l) - \Sigma_{ij}(l) - \frac{\rho_i}{\rho_j}\Sigma_{ji}(l)) < 0$$

which yields

$$\Xi_{11}^{ij}(l) + \Xi_{11}^{ji}(l) - \Sigma_{ij}(l) - \frac{\rho_i}{\rho_j}\Sigma_{ji}(l) < 0 \tag{10.32}$$

From (10.26), (10.27) and (10.32), we have

$$\dot{V}(t, x_t) \leq -z^T(t)z(t) + \gamma^2 \omega^T(t)\omega(t) \tag{10.33}$$

Since $\bigcup_{k=0}^{+\infty} \bigcup_{l=0}^{d} \Omega_{l.k} = [0, +\infty)$ and $V(t, x_t)$ is continuous in t, the integration of both sides of (10.33) from 0 to $+\infty$ yields

$$V(+\infty, x_{+\infty}) - V(0, x_0) \leq \int_0^{+\infty} [-z^T(t)z(t) + \gamma^2 \omega^T(t)\omega(t)]dt \tag{10.34}$$

Under zero initial condition $V(0) = 0$ and from (10.34), we have

$$\int_0^{+\infty} z^T(t)z(t)dt \leq \int_0^{+\infty} \gamma^2 \omega^T(t)\omega(t))dt \tag{10.35}$$

that is, $\|z(t)\| \leq \gamma \|\omega(t)\|$ for any nonzero $\omega(t) \in L_2[0, +\infty)$. With a condition that $\omega(t) = 0$, there exists a scalar $\varepsilon > 0$ such that $\dot{V}(t) \leq -\varepsilon \|x(t)\|^2$ for $x(t) \neq 0$, from which we can confirm the system (10.16) is asymptotically stable with an H_∞ performance γ. This completes the proof. $\square$

Based on Theorem 10.1, we are in a position to design the PDC fuzzy controller.

Theorem 10.2 *For some given positive constants $\bar{\eta}$, γ, λ_1 and λ_2, under the communication scheme (10.6), the system (10.16) is asymptotically stable with an H_∞ norm bound γ and controll feedback gain $K_j = Y_j X^{-1}$, if there exist real matrices $T > 0$, $P_i > 0$, $R_i > 0$ $(i = 1, 2)$ and matrices M, N, and L with appropriate dimensions such that the following matrix inequalities hold for $k, i, j = 1, 2, \ldots, r$, $i < j, l = 1, 2$:*

$$
\begin{bmatrix} \tilde{\Xi}_{11}^{kk} & * \\ \tilde{\Xi}_{21}^{kk}(l) & \tilde{\Xi}_{22}(l) \end{bmatrix} < 0 \tag{10.36}
$$

$$
\begin{bmatrix} \tilde{\Xi}_{11}^{ij} + \lambda_1 \tilde{\Xi}_{11}^{ji} & * & * \\ \tilde{\Xi}_{21}^{ij}(l) & \tilde{\Xi}_{22}(l) & * \\ \sqrt{\lambda_2}\tilde{\Xi}_{21}^{ji}(l) & 0 & \tilde{\Xi}_{22}(l) \end{bmatrix} < 0 \tag{10.37}
$$

$$
\begin{bmatrix} \tilde{\Xi}_{11}^{ij} + \lambda_2 \tilde{\Xi}_{11}^{ji} & * & * \\ \tilde{\Xi}_{21}^{ij}(l) & \tilde{\Xi}_{22}(l) & * \\ \sqrt{\lambda_2}\tilde{\Xi}_{21}^{ji}(l) & 0 & \tilde{\Xi}_{22}(l) \end{bmatrix} < 0 \tag{10.38}
$$

where

$$
\tilde{\Xi}_{11}^{ij} = \tilde{\Gamma}^{ij} + \tilde{\Delta} + \tilde{\Delta}^T
$$
$$
\tilde{\Xi}_{21}^{ij}(1) = col\{\bar{\eta}\tilde{F}_2^{ij}, \bar{\eta}\tilde{F}_2^{ij}, \bar{\eta}\tilde{L}^T, \tilde{F}_3^{ij}, \tilde{F}_2^{ij}, \bar{\eta}\tilde{R}_1 F_1\}
$$
$$
\tilde{\Xi}_{21}^{ij}(2) = col\{\bar{\eta}\tilde{F}_2^{ij}, \bar{\eta}\tilde{N}^T, \bar{\eta}\tilde{M}^T, \tilde{F}_3^{ij}\}
$$
$$
\tilde{\Xi}_{22}(1) = -diag\{\bar{\eta}X\tilde{T}^{-1}X, \bar{\eta}X\tilde{R}_2^{-1}X, \bar{\eta}\tilde{T}, I, \varepsilon I, X\varepsilon^{-1}X\}
$$
$$
\tilde{\Xi}_{22}(2) = -diag\{\bar{\eta}X\tilde{T}^{-1}X, \bar{\eta}\tilde{R}_2, \bar{\eta}\tilde{T}, I\}
$$

with

$$
\tilde{\Gamma}^{ij} = \begin{bmatrix} A_i X + X A_i^T + \tilde{P}_2 - \tilde{R}_1 & * & * & * & * & * \\ Y_j^T B_i^T & \delta\tilde{\Phi} & * & * & * & * \\ -Y_j^T B_i^T & 0 & -\tilde{\Phi} & * & * & * \\ 0 & 0 & 0 & -\tilde{P}_2 & * & * \\ \tilde{R}_1 & 0 & 0 & 0 & -\tilde{R}_1 & * \\ B_{\omega i}^T & 0 & 0 & 0 & 0 & -\gamma^2 \end{bmatrix}
$$
$$
\tilde{\Delta} = [\tilde{M} + \tilde{N}, -\tilde{M} + \tilde{L}, 0, -\tilde{L}, -\tilde{N}, 0]
$$
$$
F_1 = [I, 0, 0, 0, -I, 0]
$$
$$
\tilde{F}_2^{ij} = [A_i X, B_i Y_j, -B_i Y_j, 0, 0, B_{\omega i}]
$$
$$
\tilde{F}_3^{ij} = [C_i X, D_i Y_j, -D_i Y_j, 0, 0, 0]
$$

Proof Define $X = P_1^{-1}$, $XP_2X = \tilde{P}_2$, $XR_1X = \tilde{R}_1$, $XR_2X = \tilde{R}_2$, $XTX = \tilde{T}$, $X\Phi X = \tilde{\Phi}$, $\mathcal{J}LX = \tilde{L}$, $\mathcal{J}MX = \tilde{M}$, $\mathcal{J}NX = \tilde{N}$, and $Y_j = KX^T$, where $\mathcal{J} = diag\{X, X, X, X, X, I\}$. Then when $l = 1$, pre- and post-multiplying both sides of (10.17), (10.18), (10.19) with $diag\{\mathcal{J}_1, \mathcal{J}_2\}$, $diag\{\mathcal{J}_1, \mathcal{J}_2, \mathcal{J}_2\}$, $diag\{\mathcal{J}_1, \mathcal{J}_2, \mathcal{J}_2\}$ and their transposes; when $l = 2$, pre- and post-multiplying both sides of (10.17), (10.18), (10.19) with $diag\{\mathcal{J}_1, \mathcal{J}_3\}$, $diag\{\mathcal{J}_1, \mathcal{J}_3, \mathcal{J}_3\}$, $diag\{\mathcal{J}_1, \mathcal{J}_3, \mathcal{J}_3\}$ and their transposes, respectively, where $\mathcal{J}_1 = diag\{X, X, X, X, X, I\}$, $\mathcal{J}_2 = diag\{T^{-1}, R_2^{-1}, X, I\}$ and $\mathcal{J}_3 = diag\{T^{-1}, X, X, I\}$. For any $\varepsilon > 0$, it follows that

$$\bar{\eta} F_1^T X R_1 \tilde{F}_2^{ij} + \bar{\eta}(\tilde{F}_2^{ij})^T R_1 X F_1 \le (\tilde{F}_2^{ij})^T \varepsilon^{-1} \tilde{F}_2^{ij} + \bar{\eta}^2 F_1^T \tilde{R}_1 X^{-1} \varepsilon X^{-1} \tilde{R}_1 F_1 \tag{10.39}$$

Using the Schur complement and based on (10.39), one can readily arrive at (10.36), (10.37), and (10.38) from (10.17), (10.18), and (10.19), respectively. This completes the proof. □

Notice that the matrix inequalities in Theorem (10.1) cannot be directly solved by Matlab LMI Toolbox because of the nonlinear items $X\Phi X$, $X\tilde{T}^{-1}X$ and $X\tilde{R}_2^{-1}X$. However, instead of the original nonconvex minimization problem, the nonlinear minimization based on the cone complementary linearization algorithm can be adopted to solve this nonconvex problem [96]. Since the detailed discussions of this algorithm can be seen in [43, 44, 96], it is omitted in this chapter.

10.4 Numerical Examples

In this section, two examples are presented to show the effectiveness of the proposed event-triggered communication scheme [16]. For comparisons with some existing results, we study the networked T-S fuzzy system (10.7) with fuzzy rules investigated in recent publications [43, 44, 47, 48].

Example 1: Consider the system (10.16) with the following parameters

$$A_1 = \begin{bmatrix} -2 & 0 \\ 0 & -0.9 \end{bmatrix}, A_2 = \begin{bmatrix} -1 & 0.5 \\ 0 & -1 \end{bmatrix}, B_1 = \begin{bmatrix} -1 & 0 \\ -1 & -1 \end{bmatrix}$$

$$B_2 = \begin{bmatrix} -1 & 0 \\ 0.1 & -1 \end{bmatrix}, K_i = \begin{bmatrix} 1 & 0 \\ 0 & 1 \end{bmatrix}, \quad i = 1, 2$$

and the corresponding membership functions for rules 1 and 2 are

$$\mu_1(z(t)) = \frac{1}{1 + e^{-2x_1(t)}}, \quad \mu_2(z(t)) = 1 - \mu_1(x_1(t)) \tag{10.40}$$

Applying Theorem 10.1, we have $\bar{\eta} = 0.91\,\text{s}$ for given triggered parameters $\delta = 0.05$ and $\Phi = 0.01I$. For the purpose of showing the improvement of the proposed event-triggered scheme, Table 10.1 lists the average transmission periods

Table 10.1 Average transmission periods by different schemes with $\bar{\tau} = 0.6\,\mathrm{s}$

Simulation time (s)	5	10	15	20	25
A time-triggered scheme in [132]	0.57	0.57	0.57	0.57	0.57
A time-triggered scheme in [47]	0.78	0.78	0.78	0.78	0.78
An event-triggered scheme in (10.6)	0.83	1.42	1.66	1.81	1.92
[47] improved by (%)	6	82	112	132	146

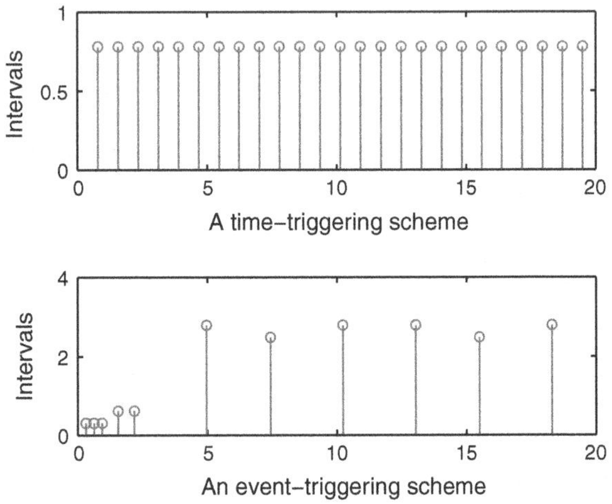

Fig. 10.1 Communication conditions in time and event-riggered schemes (Example 1)

A time–triggering scheme

An event–triggering scheme

generated by different transmission schemes and simulation times (in the simulation, h is set as $0.3\,\mathrm{s}$). It is seen from Table 10.1 that the time-triggered transmission scheme leads to overprovisioning communication bandwidth since the same communication periods are used for data transmission regardless of the behavior of the system being controlled. In particular, for the time-triggered scheme in [47, 132], when the system trends to its equilibrium, a large number of unnecessary data are still transmitted through the communication network. Moreover, it is seen that the transmission intervals generated by the event-triggered scheme in this chapter are state-dependent and time-varying. As a result, longer average communication periods are generated, such as, compared with 0.57 and $0.78\,\mathrm{s}$ in [132] and [47], 1.81 and $1.92\,\mathrm{s}$ are obtained for the simulation times $T = 20\,\mathrm{s}$ and $T = 25\,\mathrm{s}$, respectively, in this chapter.

In the case of $T = 20\,\mathrm{s}$ in Table 10.1, Fig. 10.1 shows the communication instants and communication intervals for the time-triggered scheme in [47] and the event-triggered communication scheme in this chapter. Compared with all sampled data are transmitted in [47], it is shown that only part of sampled data are transmitted in this chapter, such as, only 17 % of sampled data are sent through the communication networks, which are very beneficial for saving limited network bandwidth.

Example 2: Consider the following mass-spring system

$$\dot{x}_1 = x_2$$
$$\dot{x}_2 = -0.01x_1 - 0.67x_1^3 + \omega + u \qquad (10.41)$$

where $x_1 \in [-1, 1]$ and $\omega = 0.2 \sin(2\pi t) \exp(-t)$ which is an external disturbance. Choose fuzzy membership functions $\mu_1(x_1) = 1 - x_1^2$ and $\mu_2(x_1) = 1 - \mu_1(x_1)$ as those in [48]. The nonlinear system (10.41) can be represented by the following T-S fuzzy model:

Rule i: If x_1 is W_i ($i = 1, 2$), then

$$\dot{x}(t) = A_i x(t) + B_i u(t) + B_{\omega i} \omega(t)$$
$$z(t) = C_i x(t) + D_i u(t) \qquad (10.42)$$

where

$$A_1 = \begin{bmatrix} 0 & 1 \\ -0.01 & 0 \end{bmatrix}, A_2 = \begin{bmatrix} 0 & 1 \\ -0.68 & 0 \end{bmatrix}, B_i = \begin{bmatrix} 0 \\ 1 \end{bmatrix},$$
$$B_{\omega i} = \begin{bmatrix} 1 \\ 0 \end{bmatrix}, C_i = \begin{bmatrix} 1 & 0 \end{bmatrix}, D_i = 0, i = 1, 2$$

For illustrating the effectiveness of the proposed communication scheme, in the following, we show the simulation results with time-triggered and event-triggered communication schemes under the same initial value $x(0) = [1, -1]^T$.

Case 10.1 General time-triggered communication scheme. Considering the system (10.16) with asynchronous premise constraints, this is out of the field of the paper in [47]. Combining the MATLAB/LMI Toolbox with the CCL algorithm [96] to solve the non-LMIs in Theorem 10.2 with $\delta = 0.8$, $\tau_0 = 0.01$, $\tau_1 = 0.4$ and different λ_1 and λ_2, as listed in Table 10.2, we have the corresponding maximum allowable lower bound γ and the controller gains K_1, K_2 for asynchronous parallel distribution compensation and synchronous parallel distribution compensation methods,

Table 10.2 Lower bound γ and controller feedback gains for different values of λ_1, λ_2 and given $\delta = 0.0001$, $\eta = 0.2$

λ_1, λ_2	γ	$(K_1^T, K_2^T)^T$
$-, -$	1.40	$\begin{pmatrix} -1.5537 & -2.4694 \\ -1.5537 & -2.4694 \end{pmatrix}$
0.9, 1.1	1.29	$\begin{pmatrix} -1.5521 & -2.3338 \\ -1.3415 & -2.3984 \end{pmatrix}$
0.95, 1.05	1.20	$\begin{pmatrix} -1.6586 & -2.3973 \\ -1.3503 & -2.4215 \end{pmatrix}$
1, 1	1.12	$\begin{pmatrix} -1.7771 & -2.4520 \\ -1.3577 & -2.4365 \end{pmatrix}$

respectively. One can see from Table 10.2 that, the obtained maximum allowable bound τ_2 from the synchronous parallel distribution compensation method is larger than that obtained from the asynchronous parallel distribution compensation method, and is smaller than that obtained without considering the effect of a communication network (that is, $\lambda_1 = \lambda_2 = 1$ in Table 10.2). Moreover, one can see from Table 10.2 that the identical controller gains K_1 and K_2 are obtained based on the asynchronous parallel distribution compensation method. In this case, the asynchronous parallel distribution compensation controller equals to a linear controller. Furthermore, when $\lambda_1 = \lambda_2 = 1$, we have $\mu_j(\theta(t)) = \mu_j(\theta(t_k))$. This implies that no communication delay is considered in the communications, that is, system (10.16) is simplified as a point-to-point connected system with an event-triggered communication scheme.

Case 10.2 Event-triggered communication scheme. For given $\delta = 0.3$, $\gamma = 3$ and $\lambda_1 = \lambda_2 = 1$, we obtain $\bar{\eta} = 0.03$, $K_1 = \begin{bmatrix} -1.2168 & -2.5566 \end{bmatrix}$, $K_2 = \begin{bmatrix} -0.74641 & -2.5556 \end{bmatrix}$ and $\Phi = \begin{bmatrix} 2.9109 & 4.6231 \\ 4.6231 & 12.175 \end{bmatrix}$, respectively. Choose $\delta = 0.1$ and 0.03 s, the state trajectories of (10.41) and communication instants and communication intervals are shown in Fig. 10.2. The system can be stabilized well while preserving the desired control performance. Moreover, the effect of the proposed communication scheme is shown clearly in these figures: The proposed event-triggered communication scheme generates a longer average communication period than a time-triggered periodic communication scheme. Such as, for the case in Fig. 10.2 ($h = 0.03$ s), the average communication period is 0.416 s, and only about 7.21% sampled data need to be transmitted by the communication networks.

Moreover, with given simulation time $T = 10$ s, and $\delta = 0.001$, 0.01 and 0.1, respectively, and different sampling period h, Table 10.3 shows that the number of packets need to be transmitted by the time-triggered communication scheme in [44, 47] and event-triggered communication scheme in this chapter. Since our

Fig. 10.2 Communication and states with $h = 0.03$ s (Example 2)

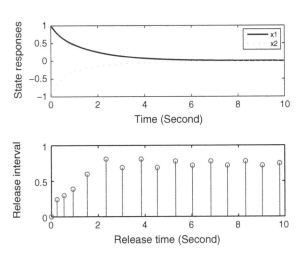

Table 10.3 Number of transmitted packets in time and event-triggered communication schemes and $T = 10\,s$

Sampling period (s)	0.01	0.02	0.03
A time-triggered scheme in [44, 47]	1000	500	333
An event-triggered scheme in (10.6) for $\delta = 0.01$	72	70	64
An event-triggered scheme in (10.6) for $\delta = 0.1$	25	25	24

approach makes a dynamic utilization of the communication bandwidth related with the current states, therefore, when the system trends to its equilibrium, the number of packets need to be transmitted reduces sharply, which indicates that the communication resource is greatly saved by the proposed communication scheme. Furthermore, one can see from Table 10.3 that the larger δ, the less number of packets need to be transmitted, which suggests a scheduling method to balance required communication bandwidth and desired control performance, that is, lowering the desired control performance allows the network to allocate more resources to other communication tasks in need.

10.5 Conclusion

A discrete event-triggered communication scheme is adopted for the communications in networked T-S fuzzy systems with asynchronous premise constraints. Compared with a time-triggered communication scheme, fewer communication resources are utilized while preserving the desired control performance. Since the proposed communication scheme only depends on the state error between the current sampled instant and the latest transmitted instant, the hardware-dependent continuous measurement of the state has been no longer needed. Moreover, by making use of the method to reconstruct the grades of membership at the controller, the same timescales of the grades of membership at the fuzzy plant and at the fuzzy controller are guaranteed. Therefore, the conservativeness induced by the asynchronous premise is reduced. Based on the proposed networked T-S fuzzy model and the discrete event-triggered communication scheme, the stabilization criterion is obtained in terms of matrix inequalities, the maximum allowable delay and the feedback gain can be obtained simultaneously through solving an optimization problem. Simulation results also show that lowering the desired control performance allows the network to allocate more resources to other communication tasks in need.

References

1. Hespanha J, Naghshtabrizi P, Xu YG (2007) A survey of recent results in networked control systems. Proc IEEE 95(1):138–162
2. Yang TC (2006) Networked control system: a brief survey. IET Control Theory Appl 153(4):403–412
3. Gupta R, Chow M (2010) Networked control system: overview and research trends. IEEE Trans Ind Electr 57(7):2527–2535
4. Zhang L, Gao H, Kaynak O (2013) Network-induced constraints in networked control systemsla survey. IEEE Trans Ind Inf 9(1):403–416
5. Sadjadpour HR, Wang Z, Garcia-Luna-Aceves J (2010) The capacity of wireless ad hoc networks with multi-packet reception. IEEE Trans Commun 58(2):600–610
6. Li HB, Yang HJ, Sun FC, Xia YQ (2014) Sliding mode predictive control of networked control systems under a multiple-packet transmission policy. IEEE Trans Ind Electr 61(11): 6234–6243
7. Isshill H, Basar T (2005) Remote control of lti systems over networks with state quantization. Syst Control Lett 54:15–31
8. Yue D, Peng C, Tang GY (2006) Guaranteed cost control of linear systems over networks with state and input quantisations. IET Control Theory Appl 153(6):658–664
9. Peng C, Tian Y-C (2007) Networked H_∞ control of linear systems with state quantization. Inf Sci 177:5763–5774
10. Wang X, Lemmon M (2011) Event-triggering in distributed networked control systems. IEEE Trans Autom Control 56(3):586–601
11. Chen G, Lewis F (2011) Distributed adaptive tracking control for synchronization of unknown networked lagrangian systems. IEEE Trans Syst Man Cybern B Cybern 99:1–12
12. Yue D, Han Q-L, Peng C (2004) State feedback controller design of networked control systems. IEEE Trans Circuits Syst II Express Briefs 51:640–644
13. Yue D, Han Q-L, Lam J (2005) Network-based robust H_∞ control of systems with uncertainty. Automatica 41(6):999–1007
14. Tian EG, Yue D, Peng C (2008) Quantized output feedback control for networked control systems. Inf Sci 178(12):2734–2749
15. Peng C, Tian Y-C, Tadé MO (2008) State feedback controller design of networked control systems with interval time-varying delay and nonlinearity. Int J Robust Nonlinear Control 18(12):1285–1301
16. Peng C, Han Q-L, Yue D (2013) To transmit or not to transmit: a discrete event-triggered communication scheme for networked takagi-sugeno fuzzy systems. IEEE Trans Fuzzy Syst 21(1):164–170

© Springer-Verlag Berlin Heidelberg 2015
C. Peng et al., *Communication and Control for Networked Complex Systems*,
DOI 10.1007/978-3-662-46813-5

17. Nilsson J et al. (1998) Real-time control systems with delays. PhD dissertation, Lund Institute of Technology, Lund
18. Quevedo DE, Nešić D (2012) Robust stability of packetized predictive control of nonlinear systems with disturbances and markovian packet losses. Automatica 48(8):1803–1811
19. Shi Y, Yu B (2009) Output feedback stabilization of networked control systems with random delays modeled by markov chains. IEEE Trans Autom Control 54(7):1668–1674
20. Nilsson J, Bernhardsson B, Wittenmark B (1998) Stochastic analysis and control of real-time systems with random time delays. Automatica 34:57–64
21. Liu M, Ho DW, Niu Y (2009) Stabilization of markovian jump linear system over networks with random communication delay. Automatica 45(2):416–421
22. Zhang J, Lam J, Xia Y (2014) Output feedback delay compensation control for networked control systems with random delays. Inf Sci 265:154–166
23. Huang D, Nguang SK (2009) Robust control for uncertain networked control systems with random delays. Springer, Berlin
24. Lin H, Antsaklis PJ (2009) Stability and stabilizability of switched linear systems: a survey of recent results. IEEE Trans Autom control 54(2):308–322
25. Zhang W, Yu L (2007) Output feedback stabilization of networked control systems with packet dropouts. IEEE Trans Autom Control 52(9):1705–1710
26. Donkers M, Heemels W, Van De Wouw N, Hetel L (2011) Stability analysis of networked control systems using a switched linear systems approach. IEEE Trans Autom Control 56(9):2101–2115
27. Liberzon D, Morse AS (1999) Basic problems in stability and design of switched systems. IEEE Control Syst 19(5):59–70
28. Lin H, Antsaklis PJ (2005) Stability and persistent disturbance attenuation properties for a class of networked control systems: switched system approach. Int J Control 78(18):1447–1458
29. Dong H, Wang Z, Ho DW, Gao H (2011) Robust filtering for markovian jump systems with randomly occurring nonlinearities and sensor saturation: the finite-horizon case. IEEE Trans Sign Process 59(7):3048–3057
30. Lin H, Zhai G, Antsaklis PJ (2003) Robust stability and disturbance attenuation analysis of a class of networked control systems. In: Proceedings of 42nd IEEE conference on decision and control, pp 1182–1187
31. Zhang WA, Yu L (2010) Stabilization of sampled-data control systems with control inputs missing. IEEE Trans Autom Control 55(2):447–610
32. Liu G-P, Xia Y, Rees D, Hu W (2007) Design and stability criteria of networked predictive control systems with random network delay in the feedback channel. IEEE Trans Syst Man Cybern Part C Appl Rev 37(2):173–184
33. Liu G-P, Xia Y, Chen J, Rees D, Hu W (2007) Networked predictive control of systems with random network delays in both forward and feedback channels. IEEE Trans Ind Electr 54(3):1282–1297
34. Xia Y, Liu G-P, Fu M, Rees D (2009) Predictive control of networked systems with random delay and data dropout. IET Control Theory Appl 3(11):1476–1486
35. Irwin G, Chen J, McKernan A, Scanlon W (2010) Co-design of predictive controllers for wireless network control. IET Control Theory Appl 4(2):186–196
36. Tian G, Xia F, Tian Y (2012) Predictive compensation for variable network delays and packet losses in networked control systems. Comput Chem Eng 39(6):152–162
37. Ding B (2011) Stabilization of linear systems over networks with bounded packet loss and its use in model predictive control. Automatica 47(11):2526–2533
38. Quevedo D, Nesic D (2011) Input-to-state stability of packetized predictive control over unreliable networks affected by packet-dropouts. IEEE Trans Autom Control 56(2):370–375
39. Montestruque LA, Antsaklis PJ (2003) On the model-based control of networked systems. Automatica 39(10):1837–1843
40. Montestruque LA, Antsaklis P (2004) Stability of model-based networked control systems with time-varying transmission times. IEEE Trans Autom Control 49(9):1562–1572

41. Feng G (2006) A survey on analysis and design of model-based fuzzy control systems. IEEE Trans Fuzzy syst 14(5):676–697
42. Tanaka K, Sano M (1994) A robust stabilization problem of fuzzy control systems and its application to backing up control of a truck-trailer. IEEE Trans Fuzzy Syst 2(2):119–134
43. Peng C, Yang TC (2010) Communication delay distribution dependent networked control for a class of T-S fuzzy system. IEEE Trans Fuzzy Syst 18(2):326–335
44. Jiang X, Han Q-L (2008) On designing fuzzy controllers for a class of nonlinear networked control systems. IEEE Trans Fuzzy Syst 16(4):1050–1060
45. Zhang HG, Yang DD, Chai TY (2007) Guaranteed cost networked control for T-S fuzzy systems with time delays. IEEE Trans Syst Man Cybern Part C Appl Rev 37(2):160–172
46. Jia X, Zhang DW, Hao XH, Zheng N (2009) Fuzzy H_∞ tracking control for nonlinear networked control systems in T-S model. IEEE Trans Syst Man Cybern B Cybern 39(4): 1073–1079
47. Tian E, Yue D, Gu Z (2010) Robust H_∞ control for nonlinear system over network: a piecewise analysis method. Fuzzy Sets Syst 161(21):2731–2745
48. Zhang HG, Yang J, Su C (2007) T-S fuzzy-model-based robust H_∞ design for networked control systems with uncertainties. IEEE Trans Ind Inf 3(4):289–301
49. Kim DW, Lee HJ (2009) Comments on "T-S fuzzy-model-based robust H_∞ design for networked control systems with uncertainties". IEEE Trans Ind Inf 5(4):507
50. Nguang SK (2010) Comment on "fuzzy H_∞ tracking control for nonlinear networked control systems in T-S model". IEEE Trans Syst Man Cybern B Cybern 40(3):957–957
51. Peng C, Yue D, Fei M (2014) Relaxed stability and stabilization conditions of networked fuzzy control systems subject to asynchronous grades of membership. IEEE Trans Fuzzy Syst 22(5):1101–1112
52. Jia X-C, Chi X-B, Han Q-L, Zheng N-N (2014) Event-triggered fuzzy H_∞ control for a class of nonlinear networked control systems using the deviation bounds of asynchronous normalized membership functions. Inf Sci 259:100–117
53. Lemmon M, Chantem T, Hu X, Zyskowski M (2007) On self-triggered full-information H_∞ controllers. In: Proceedings of hybrid system: computation control conference, pp 371–384
54. Anta A, Tabuada P (2008) Self-triggered stabilization of homogeneous control systems. In: Proceedings of American control conference, pp 4129–4134
55. Hristu-Varsakelis D, Kumar P (2002) Interrupt-based feedback control over a shared communication medium. In: Proceedings of 41st IEEE conference decision control, pp 3223–3228
56. Yook J, Tilbury D, Soparkar N (2002) Trading computation for bandwidth: reducing communication in distributed control systems using state estimators. IEEE Trans Control Syst Technol 10(4):503–518
57. Xu Y, Hespanha J (2004) Optimal communication logics in networked control systems. In: Proceedings of 43rd IEEE conference on decision and control, vol 4, pp 3527–3532
58. Xu Y, Hespanha J (2006) Communication logic design and analysis for networked control systems. In: Menini L, Zaccarian L, Abdallah CT (eds) Current trends in nonlinear systems and control. Birksäusker, Boston
59. Tabuada P (2007) Event-triggered real-time scheduling of stabilizing control tasks. IEEE Trans Autom Control 52(9):1680–1685
60. Lunze J, Lehmann D (2010) A state-feedback approach to event-based control. Automatica 46:211–215
61. Zhang JH, Feng G (2014) Event-driven observer-based output feedback control for linear systems. Automatica 50:1852–1859
62. Zhang X-M, Han Q-L (2013) Event-triggered dynamic output feedback control for networked control systems. IET Control Theory Appl 8(4):226–234
63. Donkers M, Heemels W (2012) Output-based event-triggered control with guaranteed L_∞ gain and improved and decentralised event-triggering. IEEE Trans Autom Control 57(6):1362–1376
64. Wang X, Lemmon MD (2009) Self-triggered feedback control systems with finite-gain L_2 stability. IEEE Trans Autom Control 54(3):452–467

65. Yue D, Tian E, Han Q-L (2011) A delay system method to design of event-triggered control of networked control systems. In: Proceedings of 50th IEEE conference on decision control and European control conference, pp 1668–1673

66. Heemels W, Donkers M, Teel A (2011) Periodic event-triggered control based on state feedback. In: 50th IEEE conference on decision and control and Eurpoean control conference (CDC-ECC), Orlando, pp 2571–2576

67. Heemels W, Donkers M, Teel A (2013) Periodic event-triggered control for linear systems. IEEE Trans Autom Control 58(4):847–861

68. Yue D, Tian E, Han Q-L (2013) A delay system method for designing event-triggered controllers of networked control systems. IEEE Trans Autom Control 58(2):475–481

69. Peng C, Yang T (2013) Event-triggered communication and H_∞ control co-design for networked control systems. Automatica 49(5):1326–1332

70. Anta A, Tabuada P (2010) To sample or not to sample: self-triggered control for nonlinear systems. IEEE Trans Autom Control 55(9):2030–2042

71. Mazo M, Tabuada P (2011) Decentralized event-triggered control over wireless sensor/actuator networks. IEEE Trans Autom Control 56(10):2456–2461

72. Guan YP, Han Q-L, Peng C (2013) Decentralized event-triggered control for sampled-data systems with asynchronous sampling. In: America control conference (ACC), pp 6565–6570

73. Yin X, Yue D, Hu S (2013) Distributed event-triggered control of discrete-time heterogeneous multi-agent systems. J Frankl Inst 350(3):651–669

74. Velasco M, Fuertes J, Martí P (2003) The self triggered task model for real-time control systems. In: Work-in-progress of the 24th IEEE real-time systems symposium (RTSS), pp 67–70

75. Tiberi U, Fischione C, Johansson K, Benedetto M (2013) Energy-efficient sampling of networked control systems over ieee 802.15. 4 wireless networks. Automatica 49:712–724

76. Mazo M, Tabuada P (2008) On event-triggered and self-triggered control over sensor/actuator networks. In: 47th IEEE conference on decision and control. IEEE, pp 435–440

77. Nowzari C, Cortés J (2012) Self-triggered coordination of robotic networks for optimal deployment. Automatica 48:1077–1087

78. Wang X, Lemmon M (2010) Self-triggering under state-independent disturbances. IEEE Trans Autom Control 55(6):1494–1500

79. Peng C, Han Q-L (2012) A novel self-trigged sampling scheme in networked control systems. In: 51st IEEE conference on decision and control, pp 3904–3909

80. Mazo Jr M, Tabuada P (2009) Input-to-state stability of self-triggered control systems. In: 48th IEEE conference on decision and control, pp 928–933

81. McKernan A, Irwin G (2010) Event-based sampling for wireless network control systems with QoS. In: American control conference (ACC), pp 1841–1846

82. Tipsuwan Y, Chow MY (2004) Gain scheduling middleware: a methodology to enable existing controllers for networked control and teleoperation-part i: network control. IEEE Trans Ind Electr 51(6):1218–1227

83. Tian Y-C, Levy D, Tadé MO, Gu T, Fidge C (2006) Queuing packets in communication networks networked control systems. In: Proceedings of the 6th world congress on intelligent control and automation (WCICA'06), Dalian, pp 210–214

84. Tian Y-C, Yu Z-G, Fidge C (2007) Multifractal nature of network induced time delay in networked control systems. Phys Lett A 361:103–107

85. Peng C, Yue D, Tian EG, Gu Z (2009) A delay distribution based stability analysis and synthesis approach for networked control systems. J Frankl Inst 346(4):349–365

86. Peng C, Tian Y, Yue D (2010) Delay distribution based robust H_∞ control of networked control systems with uncertainties. Asian J Control 12(1):46–57

87. Peng C, Han Q-L (2013) A novel event-triggered transmission scheme and L_2 control co-design for sampled-data control systems. IEEE Trans Autom Control 58(10):2620–2626

88. Peng C, Tian Y-C, Yue D (2011) Output feedback control of discrete-time systems in networked environments. IEEE Trans Syst Man Cybern Part A Syst Hum 41(1):185–190

89. Yang F, Wang Z, Hung Y, Gani M (2006) H_∞ control for networked systems with random communication delays. IEEE Trans Autom Control 51(3):511–518

90. Hu SL, Yue D (2013) L_2-gain analysis of event-triggered networked control systems: a discontinuous lyapunov functional approach. Int J Robust Nonlinear Control 23(11):1277–1300

91. Gu K, Kharitonov VL, Chen J (2003) Stability of time-delay systems. Birkhauser, Boston

92. Park P, Ko J, Jeong C (2011) Reciprocally convex approach to stability of systems with time-varying delays. Automatica 47(1):235–238

93. Peng C, Fei MR (2013) An improved result on the stability of uncertain T-S fuzzy systems with interval time-varying delay. Fuzzy Sets Syst 212:97–109

94. Yue D, Tian E, Zhang Y, Peng C (2009) Delay-distribution-dependent stability and stabilization of T-S fuzzy systems with probabilistic interval delay. IEEE Trans Syst Man Cybern Part B Cybern 39(2):503–516

95. de Souza CE, Li X (1999) Delay-dependent robust H_∞ control of uncertain linear state-delayed systems. Automatica 35:1313–1321

96. Ghaoui LE, Oustry F, AitRami M (1997) A cone complementarity linearization algorithm for static output-feedback and related problems. IEEE Trans Autom Control 42(8):1171–1176

97. Park H, Kim Y, Kim D, Kwon W (2002) A scheduling method for network based control systems. IEEE Trans Control Syst Technol 10:318–330

98. Kim DS, Lee YS, Kwon WH, Park HS (2003) Maximum allowable delay bounds of networked control systems. Control Eng Pract 11:1301–1313

99. Jiang X, Han Q-L (2006) Delay-dependent robust stability for uncertain linear systems with interval time-varying delay. Automatica 42:1059–1065

100. Walsh GC, Ye H, Bushnell LG (2002) Stability analysis of networked control systems. IEEE Trans Control Syst Technol 10:438–446

101. Takagi T, Sugeno M (1995) Fuzzy identification of systems and its applications to modeling and control. IEEE Trans Syst Man Cybern 1:116–132

102. Yang FW, Wang ZD, Hung YS, Gani M (2006) H_∞ control for networked systems with random communications delays. IEEE Trans Autom Control 51(3):511–518

103. Zhang HG, Li M, Yang J, Yang D (2009) Fuzzy model-based robust networked control for a class of nonlinear systems. IEEE Trans Syst Man Cybern Part A Syst Hum 39(2):437–447

104. Tang W, Chen G, Lu R (2001) A modified fuzzy PI controller for a fiexible-joint robot arm with uncertainties. Fuzzy Sets Syst 118(1):109–119

105. Peng C, Han Q-L, Yue D (2013) Communication-delay-distribution dependent decentralized control for large-scale systems with IP-based communication networks. IEEE Trans Control Syst Technol 21(3):820–830

106. Ikeda M, Šiljak D (1992) Robust stabilisation of nonlinear systems via linear state feedback. In: Leondes CT (ed) Control and dynamic systems, vol 51. Academic Press, New York

107. Yang TC, Peng C, Yue D, Fei MR (2010) New study of controller design for networked control systems. IET Control Theory Appl 4(7):1109–1121

108. Gao H, Meng X, Chen T (2008) Stabilization of networked control systems with a new delay characterization. IEEE Trans Autom Control 53(9):2142–2148

109. Jiang X, Han Q-L, Liu S, Xue A (2008) A new H_∞ stabilization criterion for networked control systems. IEEE Trans Autom Control 53(4):1025–1031

110. Xiong J, Lam J (2009) Stabilization of networked control systems with a logic ZOH. IEEE Trans Autom Control 54(2):358–363

111. Johansson K (2000) The quadruple-tank process: a multivariable laboratory process with an adjustable zero. IEEE Trans Control Syst Technol 8(3):456–465

112. Labibi B, Marquez H, Chen T (2009) Decentralized robust output feedback control for control affine nonlinear interconnected systems. J Process Control 19(5):865–878

113. Zhang XM, Han Q-L (2006) Delay-dependent robust H_∞ filtering for uncertain discrete-time systems with time-varying delay based on a finite sum inequality. IEEE Trans Circuits Syst II Analog Digital Signal Process (in press)

114. Peng C, Fei MR (2013) Networked H_∞ filtering for discrete linear systems with a periodic event-triggering communication scheme. IET Signal Process 7(8):754–765

115. Peng C (2012) Improved delay-dependent stabilization criteria for discrete systems with a new finite sum inequality. IET Control Theory Appl 6(3):448–453

116. Zhang H, Shi Y, Mehr AS (2011) Robust weighted H_∞ filtering for networked systems with intermittent measurements of multiple sensors. Int J Adapt Control Signal Process 25(4): 313–330

117. Gao H, Chen T (2007) New results on stability of discrete-time systems with time-varying state delay. IEEE Trans Autom Control 52(2):328–334

118. Shao HY, Han Q-L (2011) New stability criteria for linear discrete time systems with interval-like time-varying delays. IEEE Trans Autom Control 56(3):619–625

119. Zhang BY, Xu SY, Zhou Y (2008) Improved stability criterion and its applications in delayed controller design for discrete time systems. Automatica 44:2963–2967

120. Gao H, Chen T (2007) H_∞ estimation for uncertain sysems with limited communication capacity. IEEE Trans Autom Control 52(11):2070–2084

121. Song H, Yu L, Zhang W (2011) Networked H_∞ filtering for linear discrete-time systems. Inf Sci 181:686–696

122. Peng C, Han Q-L (2013) A novel event-triggered transmission scheme and L_2 control co-design for sampled-data control systems. IEEE Trans Autom Control 58(10):2620–2626

123. He Y, Liu GP, Rees D, Wu M (2007) Improved stabilisation method for networked control systems. IET Control Theory Appl 1(6):1580–1585

124. Mazo M Jr, Anta A, Tabuada P (2010) An ISS self-triggered implementation of linear controllers. Automatica 46(8):1310–1314

125. Kwon OM, Park J, Park MJH, Lee SM, Cha EJ (2013) Improved delay-dependent stability criteria for discrete-time systems with time-varying delays. Circuits Syst Signal Process 32(4):1949–1962

126. Willig A (2008) Recent and emerging topics in wireless industrial communication. IEEE Trans Ind Inf 4(2):102–124

127. Tiberi U, Fischione C, Johansson K, Benedetto MDD (2010) Adaptive self-triggered control over IEEE 802.15. 4 networks. In: 49th IEEE conference on decision and control (CDC), pp 2099–2104

128. Yue D, Tian E, Han Q-L (2013) A delay system method for designing event-triggered controllers of networked control systems. IEEE Trans Autom Control 58(2):475–481

129. Khalil HK (2002) Nonlinear systems, 3rd edn. Pretience Hall, New Jersey

130. Anta A, Tabuada P (2008) Space-time scaling laws for self-triggered control. In: 47th IEEE conference on decision and control, pp 4420–4425

131. Zhang X, Lu G, Zheng Y (2008) Stabilization of networked stochastic time-delay fuzzy systems with data dropout. IEEE Trans Fuzzy Syst 16(3):798–807

132. Peng C, Yue D, Tian Y-C (2009) New approach on robust delay-dependent H_∞ control for uncertain T-S fuzzy systems with interval time-varying delay. IEEE Trans Fuzzy Syst 17(4):890–900

133. Zhang W, Branicky MS, Phillips SM (2001) Stability of networked control systems. IEEE Control Syst Mag 21:84–99

Index

© Springer-Verlag Berlin Heidelberg 2015
C. Peng et al., *Communication and Control for Networked Complex Systems*,
DOI 10.1007/978-3-662-46813-5

Printed in the United States
By Bookmasters